T0353516

Flash Memory Integration

To the memory of my father, to Anas and Assil,
to my family from here and there
Jalil

To my parents
Pierre

Energy Management in Embedded Systems Set

coordinated by
Maryline Chetto

Flash Memory Integration

Performance and Energy Considerations

Jalil Boukhobza
Pierre Olivier

ELSEVIER

First published 2017 in Great Britain and the United States by ISTE Press Ltd and Elsevier Ltd

ISTE Press Ltd
27-37 St George's Road
London SW19 4EU
UK

www.iste.co.uk

Elsevier Ltd
The Boulevard, Langford Lane
Kidlington, Oxford, OX5 1GB
UK

www.elsevier.com

Notices

Knowledge and best practice in this field are constantly changing. As new research and experience broaden our understanding, changes in research methods, professional practices, or medical treatment may become necessary.

Practitioners and researchers must always rely on their own experience and knowledge in evaluating and using any information, methods, compounds, or experiments described herein. In using such information or methods they should be mindful of their own safety and the safety of others, including parties for whom they have a professional responsibility.

To the fullest extent of the law, neither the Publisher nor the authors, contributors, or editors, assume any liability for any injury and/or damage to persons or property as a matter of products liability, negligence or otherwise, or from any use or operation of any methods, products, instructions, or ideas contained in the material herein.

For information on all our publications visit our website at http://store.elsevier.com/

© ISTE Press Ltd 2017
The rights of Jalil Boukhobza and Pierre Olivier to be identified as the authors of this work have been asserted by them in accordance with the Copyright, Designs and Patents Act 1988.

British Library Cataloguing-in-Publication Data
A CIP record for this book is available from the British Library
Library of Congress Cataloging in Publication Data
A catalog record for this book is available from the Library of Congress
ISBN 978-1-78548-124-6

Printed and bound in the UK and US

Contents

Acknowledgments

Acknowledgments by Jalil Boukhobza

Needless to say, the writing of this book would not have been possible without the precious help of several people. First of all, those who contributed to this book, starting with my co-author and my first PhD student Pierre Olivier, without whom I would have never been able to meet the deadlines. I also wish to thank Hamza Ouarnoughi, one of my PhD students, who has participated extensively in experiments. I thank him for his availability and for his exemplary responsiveness. I thank Jean-Émile Dartois, R&D engineer at IRT b<>com, who has also contributed a great deal to the experimental part, for his commitment and for his relevant suggestions.

I also wish to express my gratitude to those who enlightened me with their proofreading and corrections; I am thinking about my wife, Linda Gardelle-Boukhobza, about Stéphane Rubini and Pierre Ficheux. I thank Frank Singhoff and Antoine Cabot as well for their support.

Finally, I thank my colleagues from UBO Department of Computer Science and the Lab-STICC laboratory who supported me in one way or another during the writing of this book, and, in particular, those who contribute day by day to the harmony and to the pleasant atmosphere within our department.

Acknowledgments by Pierre Olivier

I thank my thesis supervisors, Jalil and Eric, for their precious advice and mentoring. I also thank all my nearest and dearest, my family and friends, who supported me during my graduate years and who still continue to do so.

Foreword 1

I have been working on *embedded Linux* systems for several years and have had the opportunity to follow Jalil and Pierre's work regarding flash memory. I have also been very pleased to do lectures at the University of Western Brittany in Brest at the request of Jalil and to attend Pierre's remarkable thesis defense.

Formerly present but unknown by the general public, today embedded systems have invaded our everyday life, and open-source software, especially the Linux operating system, plays a prominent role in these systems. The Linux kernel is present in Google's Android system, which is employed in more than a billion mobile phones. The Linux operating system (in its *embedded* version) is ubiquitous in gateways for access to the Internet and to digital television (the famous *set-top box*). Recently, the very same Linux has invaded the automotive industry and backed so-called IVI systems (In Vehicle Infotainment), which include, among others, navigation, multimedia streaming in the passenger compartment or driver assistance. Nowadays, any multimedia device which does not mention Internet of Things (IoT) seems old-fashioned. In the same way as the Internet overwhelmed our daily life in the last century, the current revolution involves integrating physical objects into a network of networks, and these physical objects are themselves embedded systems. Every object has to store at least some programs and data, and as such the importance of flash memory has become evident if we consider the exponential number of deployed systems (we are talking about tens of billions in some decades time).

Of course, great progress has being achieved, as most mobile phones have 16 GB, or even more, of flash memory, whereas in 1995 a 5 MB hard drive occupied the size of a large wardrobe! In those days, the computer was far from being a product of mass consumption, whereas nowadays a USB drive (which contains at least 8 GB of flash memory) is readily available in the supermarket or at gas stations alongside sweets, batteries and water bottles. This great progress will have (and has already had) an effect on our environment, and for this reason it is a fundamental research subject, since after all, it is mentioned rather little in the specialist literature. The fact that such a complex technical subject is being trivialized shows that a great amount of work by researchers, engineers and industrialists from different fields (electronics, software and mechanics) is required to reach a satisfactory result. Admittedly, the task is far from finished and most likely present-day products will make us laugh in a few decades time.

This book, which embodies Jalil and Pierre's thousands of hours of work and effort, provides an impressive brick to the building of technological progress. As any book produced since the dawn of time, it represents a part of world's memory and it completes in an indispensable way the instantaneous (though not always verifiable) information available on the Web.

Pierre FICHEUX
January 2017

Foreword 2

Our societies are going through a genuine digital revolution, known as the Internet of Things (IoT). This is characterized by increased connectivity between all kinds of electronic materials, from surveillance cameras to implanted medical devices. With the coming together of the worlds of computing and communication, embedded computing now stretches across all sectors, public and industrial. The IoT has begun to transform our daily life and our professional environment by enabling better medical care, better security for property and better company productivity. However, it will inevitably lead to environmental upheaval, and this is something that researchers and technologists will have to take into account in order to devise new materials and processes.

Thus, with the rise in the number of connected devices and sensors, and the increasingly large amounts of data being processed and transferred, demand for energy will also increase. However, global warming is creating an enormous amount of pressure on organizations to adopt strategies and techniques at all levels that prioritize the protection of our environment and look to find the optimal way of using the energy available on our planet.

In recent years, the main challenge facing R&D has been what we have now come to refer to as green electronics/computing: in other words, the need to promote technological solutions that are energy efficient and that respect the environment.

The following volumes of the Energy Management in Embedded Systems Set have been written in order to address this concern:

– In their volume *Energy Autonomy of Batteryless and Wireless Embedded Systems*, Jean-Marie Dilhac and Vincent Boitier consider the question of the energy autonomy of embedded electronic systems, where the classical solution of the electrochemical storage of energy is replaced by the harvesting of ambient energy. Without limiting the comprehensiveness of their work, the authors draw on their experience in the world of aeronautics in order to illustrate the concepts explored.

– The volume *ESD Protection Methodologies*, by Marise Bafleur, Fabrice Caignet and Nicolas Nolhier, puts forward a synthesis of approaches for the protection of electronic systems in relation to electronic discharges (ElectroStatic Discharge or ESD), which is one of the biggest issues with the durability and reliability of new technology. Illustrated by real case studies, the protection methodologies described highlight the benefit of a global approach, from the individual components to the system itself. The tools that are crucial for developing protective structures, including the specific techniques for electrical characterization and detecting faults as well as predictive simulation models, are also featured.

– Maryline Chetto and Audrey Queudet present a volume entitled *Energy Autonomy of Real-Time Systems*. This deals with the small, real-time, wireless sensor systems capable of taking their energy from the surrounding environment. Firstly, the volume presents a summary of the fundamentals of real-time computing. It introduces the reader to the specifics of so-called autonomous systems that must be able to dynamically adapt their energy consumption to avoid shortages, while respecting their individual time restrictions. Real-time sequencing, which is vital in this particular context, is also described.

– Finally this volume, entitled *Flash Memory Integration*, by Jalil Boukhobza and Pierre Olivier attempts to highlight what is currently the most commonly used storage technology in the field of embedded systems. It features a description of how this technology is integrated into current systems and how it acts from the point of view of performance and energy consumption. The authors also examine how the energy consumption and the performance of a system are characterized at the software level (applications, operating system) as well as the material level (flash memory, main memory and CPU).

Maryline CHETTO
January 2017

PART 1

Introduction

Yesterday is but today's memory and tomorrow is today's dream

Khalil Gibran

Memory is imagination in reverse
Daniel Pennac

General Introduction

This chapter introduces the major challenges related to data storage. These very challenges constitute the inspiration for writing this book. In this first chapter, we will provide some facts regarding the outburst of digital data and storage needs, and the importance of taking into account power consumption in information technology with a focus on different storage systems and, more generally, on memory hierarchy.

The outline of this chapter is as follows:

– the first section describes the digital *Big Data deluge* phenomenon;

– the second section underlines the importance of power consumption, related both to processing and storage of data;

– the third section describes memory hierarchy and its importance in computer systems;

– finally, the last section introduces flash memory and its integration in computer systems.

1.1. The outburst of digital data

Nowadays, the amount of generated digital data, in all its forms, is following an exponential growth. Indeed, this volume is growing faster than Moore's law (although the latter is not intended to represent such metrics). As an example, online data indexed by Google increased by a factor of 56 between 2002 and 2009 (from 5 to 280 exabytes) [RAN 11]. This trend is far from being unique to web data; the data volume of companies is undergoing this fast growth as well. A cumulative increase of 173% per year has been observed [RAN 11, WIN 08].

In a recent study [FAR 10], EMC (a major player in the data storage sector) forecasts that the volume of digital data created each year will increase by a factor of 44 between 2009 and the 2020 horizon. EMC also predicts that more than a third of this amount of data will be stored in or will transit through the Cloud. This fact demonstrates the importance of the need for additional storage capacity that can absorb this deluge of data.

In fact, software applications are processing data more and more intensively. As an example, we can mention applications for sharing video and images on social networks, online transaction processing, search engines, e-mail processing, digital television, etc. We could also talk about *Big Data*, a term which designates new techniques related to the processing of the digital data deluge (as this book does not address this aspect, the interested reader can refer to books that are dedicated to this topic). Actually, the Big Data phenomenon forces us to consider not only new methods of data processing, but also new ways of storing data and extracting them from storage media.

In a broader sense, digital data are, by their nature, heterogeneous and they are processed and managed using different techniques and methods: acquisition, analysis, processing, classification, storage, etc. [RAN 11]. This heterogeneity in processing, as well as the significant amounts involved, are factors that put more and more pressure on storage subsystems and on the entire memory hierarchy. This issue is urging the scientific community to consider new technologies and methods which could optimize the performance of these systems.

1.2. Performance and power consumption of storage systems

In addition to performance, power consumption is becoming an increasingly critical metric. In 2006, the data centers in the United States already consumed more than 1.5% of energy generated in the country, and it has been estimated that this percentage will increase by 18% per year [ZHA 10]. In 2013, a report by *Digital Power Group* revealed that information and communications technology (ICT), which comprises *smartphones*, notebooks, PCs, digital TVs, and, in particular, servers, used about 1500 terawatt-hours of electricity per year in the world. This amount is comparable to the total annual electricity produced by Japan and Germany. It is also equivalent to the total energy used in the whole world in 1985[1]. The same report estimates that ICT will consume around 10% of electricity produced worldwide at the 2020 horizon. Other figures predict 15% of world energy for ICT.

More specifically, power consumption of data center storage systems should represent between 20% and 40% of total power consumption [CAR 10]. In other studies, an order of power consumption of 14% can be found for the storage system (see Figure 1.1). As a consequence, in addition to the considerations related to the metrics of performance and storage capacity or volume, energy efficiency becomes one of the most important metrics to be taken into account in order to reduce the operating costs of a *data center* ([ROB 09]), and, in absolute terms, their energy footprint.

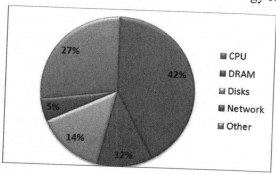

Figure 1.1. *Approximated distribution of peak power usage by hardware components in a data center that uses servers of 2012th generation [HOE 09]. For a color version of this figure, see www.iste.co.uk/boukhobza/flash.zip*

In order to partially reduce the cost of power consumption (reduction of the running cost), IT giants such as Microsoft, Google and Yahoo have tackled the construction of data centers close to power plants [VAS 10]. Indeed, data centers are guided more and more by the management of their power budget [CAR 10].

---

1 http://science.time.com/2013/08/14/power-drain-the-digital-cloud-is-using-more-energy-than-you-think/

Knowing that only a small part of world data is digitalized (about 5% [RAN 11]), the increasing requirements of storage capacity, performance and power consumption will be pursued for many years to come. Therefore, it is crucial to consider future evolution in accordance with energy metrics, both from a technological and architectural points of view.

1.3. Memory hierarchy and storage technology

Memory is a key element in a computer system, and both its function and its integration in the systems have evolved enormously over time, giving rise to multiple peripherals employing different kinds of technology. With regard to power consumption, memory (both main and secondary) represents between 25% and 40% of power consumption in a computer system. This proportion is very significant, although it is exceeded by CPU, which consumes between around 30% and 50% of energy, depending on its load. However, about half of the CPU consumption is due to the cache memory [HOR 14]. In other words, memory hierarchy, in its totality, is the element that consumes the most power in a computer system.

During the design of computer systems, memory integration has introduced early on a first kind of division in the management of memory hierarchy: (1) a fast memory that supports access by bytes and fast transfer of data, and which is essential for the execution of applications, called primary or main memory; (2) a memory used for mass storage, which completes the main memory thanks to a lower cost, called secondary memory. This creation of a memory hierarchy has led to different management policies at the operating system level. In fact, main memory's management and its sharing among different tasks of an operating system are performed by a specific service of memory management. Meanwhile the secondary memory's management is carried out by the Input/Output (I/O) devices management system, as defined by the Von Neumann architecture.

In addition to the main and secondary memory, many other types have found a place in the memory hierarchy (see Figure 1.2), always with the same primary goal: to optimize the performance and to reduce the ever-growing gap in performance between processors and memory. We are talking about cache memory of different levels. It is important to note that, although they can be deactivated (or activated) by the operating system, this cache memory is not handled by the OS; its management is performed by means of dedicated circuits.

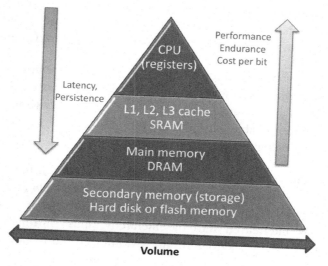

Figure 1.2. *Illustration of the memory hierarchy pyramid*

Considering the memory hierarchy, we have to reconcile three points of view related to architecture, system and application. From the architectural point of view, the position of a memory technology within the pyramid of memory hierarchy (see Figure 1.2) reflects its distance from the processor. Generally, the more significant the performance and the endurance of a memory technology, the closer this memory is situated to the processor. On the contrary, the technologies with the worst performance are situated at the bottom of the pyramid, as, for example the traditional hard disk drives, followed by the magnetic tapes.

From the system point of view, the operating system recognizes three kinds of memory: cache, main memory and secondary memory, although it handles only two levels, the main and the secondary memory. The main memory is addressed byte wise and it is managed as a shared resource by means of a dedicated service of the operating system. The operating system can employ specific equipment, such as associative memory (or TLB, for Translation Look-aside Buffer), in order to accelerate the management of these kinds of memory. Finally, the secondary memory consists of storage systems based on traditional hard disk drives or flash memory (for example, in the form of a Solid State Disk or SSD). As for the storage system, it is considered on a par with a peripheral which is accessed by blocks (for instance, a sector in a hard disk drive) and is managed as an I/O device.

Finally, from the application point of view, cache memory is entirely abstracted. Meanwhile, the main and the secondary memory are accessible via different interfaces: a rather direct access to the memory (for example, by means of variables or tables in C language), and access by means of a file system for the secondary storage.

From the point of view of the architecture, in the case of the introduction of a new memory in the memory hierarchy, we can talk about a vertical or horizontal integration. A vertical integration means that a new memory technology has interleaved into the hierarchy, thus shifting an existing technology towards the bottom of the pyramid. A horizontal integration means that the newly integrated technology is situated at the same level of an existing technology, and usually, in this case, the two technologies share the same interface. As an example, flash memory can be integrated at the same level as the hard disk drives in the memory hierarchy, in which case it can share the same interfaces, for instance, SATA interface for SSDs and hard disk drives (see Figure 1.3).

Whether the integration of a memory technology is horizontal or vertical, the hardware part (for example, the controller) and the software part (the operating system and eventually the application) have to take into account this integration for a better optimization of performance.

1.4. Introduction to flash memory integration

Flash memory is a semiconductor non-volatile EEPROM (*Electrically Erasable and Programmable Read Only Memory*) memory invented in Toshiba laboratories in the 1980s. The fundamental component of flash memory is a floating-gate transistor which ensures the non-volatility of data by maintaining a charge in the floating gate when the memory is not supplied with power (see Chapter 2).

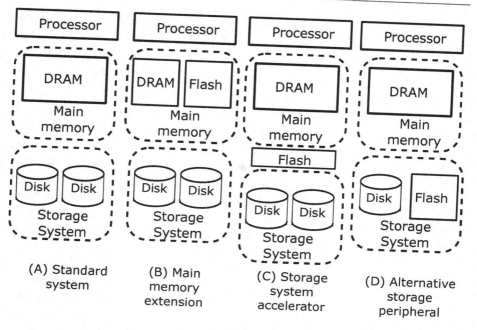

Figure 1.3. *Horizontal and vertical integration of flash memory in computer systems*

Flash memory started being used in embedded products in the 1990s, for instance by means of an SD card, CompactFlash or, otherwise, by direct integration in electronic boards. It is mainly thanks to the popularization of smartphones around 2007–2008 that this technology has become essential. In fact, mass production of these devices has allowed a large reduction of manufacturing costs. Ultimately, it is only at the beginning of the 2000s that their use has been popularized in the field of personal computers.

Flash memory has been integrated in present computer systems in three different ways [ROB 09, GUR 09, BOU 13b] (see Figure 1.3):

1) as an extension of main memory (horizontal integration), see Figure 1.3-(B);

2) as an accelerator of the storage system, in which case the flash memory is considered as a cache (with respect to the disk) which stores frequently accessed data in order to reduce the access to the disks (vertical integration), see Figure 1.3-(C);

3) as a peripheral for alternative or complementary storage to the traditional storage on the hard disk drive, generally in the form of an SSD (Solid State Drive), see Figure 1.3-(D). In this case, it is, once again, a horizontal integration in which the SSD is at the same level as the hard disk drive from the point of view of memory hierarchy.

From the three scenarios enumerated above, we can observe that there are two possibilities of horizontal integration at two different levels of memory hierarchy, i.e. in the main memory and in the secondary memory, and one vertical integration between these two levels. Actually, flash memory performance, in terms of speed and latency, is situated between those of hard disk drives and those of DRAM (Direct Random Access Memory). However, in practice, integration is seldom done at DRAM level because of the limited lifetime of flash cells. The endurance of a DRAM in terms of write cycles is of the order of 10^{15}, whereas the endurance of a flash memory is equal to 10^5 at best. The rare uses of the flash memory at the same level as the DRAM are done for fast recovery after a power failure. In fact, in this case, the DRAM module is coupled with a capacitor which discharges the content of the DRAM on a flash memory after an electric failure; this content will be recharged directly from the flash memory, thus drastically reducing the restarting time.

From the application standpoint, integration at the DRAM level is completely imperceptible, as all the elements which manage this technological heterogeneity are hardware components. In the case of a horizontal integration as a secondary storage, the flash memory is seen by the application software in the same way of traditional hard disk drives, i.e. as a block device which uses the same interfaces and possesses its own file system. Because of its performance capabilities, flash memory has been integrated in some peripherals by means of more efficient interfaces, such as PCI (Peripheral Component Interconnect). In this way, the internal parallelism available in flash peripherals can be used entirely, without limitations imposed by the performance of a classical I/O interface (for example, SATA). This kind of integration may require the development of device drivers and specific tools for reducing the latency due to the traditional software stack for I/O management.

Despite the performance that flash memory can provide, it will not replace traditional hard disk drives in the short or medium term. Two phenomena can explain this fact. On the one hand, there is the cost of flash memory, expressed in $ per gigabyte, which on average is one order of magnitude greater than that of traditional hard disk drives. On the other hand, irrespective of any cost limits, the outburst of produced digital data is more considerable than the manufacturing of flash memory. In fact, the present production of flush memory is incapable of absorbing the amount of generated data.

1.5. Scope of the book

Flash memory has invaded many of our devices at a remarkable speed, and it has changed the way we conceive the storage of our data. It has urged engineers to reconsider the way of thinking about storage in several application domains, whether in the domain of embedded devices, of high performance computing, or, else, of the Cloud. Flash memory is present in our phones, our cameras, our Internet boxes, our televisions, our cars, and also in high performance computing centers. This omnipresence can be explained by their features in terms of performance and power consumption.

The goal of this book is to help understand how flash memory works, its integration in current computer systems from multiple application domains, and finally outline flash memory behavior in terms of performance and power consumption.

1.6. Target audience

This book is addressed to computer scientists of any level who wish to broaden their knowledge or simply to discover the operating principles of flash memory. In this book, we have opted for a system point of view of flash memory, therefore making abstract the technological specifics at a very low level in order to facilitate the reading for computer scientists. Also, we have opted for an applied and experimental approach illustrating with measurements the different concepts described, relying on a GNU/Linux-based system.

The goal of this book is to allow application developers, administrators and simple users to understand the operating principles of their storage systems based on flash memory and to comprehend the differences between them and traditional storage systems based on hard disk drives. In fact, as opposed to the latter, flash memory has asymmetrical reading/writing performances and it can have a very large design space, which justifies the gap in performance and price that can be very significant, for example between two SSDs.

To resume, this book is addressed to computer scientists of any level, but it can also be read by people who are interested in this domain or by electronic engineers; it discusses flash memory, and does this for several application domains, and it contains an exhaustive experimental study performed at the laboratory Lab-STICC (CNRS UMR 6285) and the Institute of Research and Technology IRT b<>com, which illustrates the behavior of this memory.

1.7. Outline of the book

The book is divided into four parts: an introductory part, a part regarding the integration of flash memory with dedicated file systems (in embedded electronic cards), a more general part about flash memory with an FTL (Flash Translation Layer) which can be found in SSDs, SD cards or USB keys, and, finally, some perspectives concerning non-volatile memory that are being investigated, such as phase-change memory (PCM) and resistive memory (ReRAM).

– Part 1: this first part is introductory and presents the basic concepts that will be employed in the remainder of the book. Chapter 1 is a general introduction to the topic discussed in this book. Chapter 2 describes the structure and the constraints of flash memory that will be the base for Parts 2 and 3. Finally, Chapter 3 deals with the evaluation of performance and power consumption in storage systems in general, and therefore it introduces the methodological concepts that help to better comprehend the methodologies discussed in Parts 2 and 3.

– Part 2: this second part discusses flash memory integrated in computer systems by means of dedicated file systems, mainly in the domain of the embedded systems. Chapter 4 expands on this kind of integration by describing the file systems for flash memory. Chapter 5 describes a methodology of exploration of performances and power consumption in such a context, based on an embedded Linux [FIC 11], whereas Chapter 6 illustrates the results of the application of this methodology.

– Part 3: this third part discusses FTL (Flash Translation Layer) based flash memory, i.e. that which can be found in SSDs, USB keys and cards with embedded eMMC (embedded Multi-Media Controller) flash memory. This part has a structure similar to the previous one, with Chapter 7 introducing the notion of FTL and its services, Chapter 8 illustrating the methodology for the experimental study of this part, and, finally, Chapter 9 detailing the results of the experimental study.

– Part 4: this part concludes the book and contains essentially two chapters. Chapter 10 introduces the non-volatile memory of the future which is intended to be added to the memory hierarchy. This memory is very promising in terms of performance and power consumption; we will describe its main types. Finally, the last chapter concludes this book.

1.8. How to read this book

The first part of this book can be read independently of the rest of the book. It contains an introduction to the operating principles of flash memories, as well as a description of evaluation methods for this kind of memory. The second part depends on the first as it employs the basic notions of flash memory and performance measure in order to apply them to dedicated file system flash memories. The third part can be read independently of the second one, but the first part contains its prerequisites. The last part, more specifically the chapter on non-volatile memories, can be read independently of Parts 2 and 3.

2

Flash Memories: Structure and Constraints

This chapter presents the general concepts related to flash memory and introduces the notions which are necessary for the comprehension of the rest of the book. We will start with a general presentation of flash memory by introducing its functioning principles and the different flash memory types. Next, we will describe the internal structure of flash memory, as well as the operations that can be performed on a chip, before presenting the main limitations in the usage of this kind of memory. Afterwards, we will approach the general principles of solutions for circumventing these limitations, which are related to the implementation of management layers in flash storage systems. These layers can be divided into two broad categories: *Flash Translation Layers* (FTL), used in flash-based peripherals, and dedicated *Flash File Systems* (FFS). These management systems are described briefly at the end of this chapter. An entire chapter will be dedicated to each of these systems: Chapter 4 for the FFS and Chapter 7 for the FTL.

In this chapter we follow the plan listed below:

1) General presentation of flash memory.

2) Constraints and limitations.

3) General concepts of constraint management systems.

4) Implementation of constraint management systems.

2.1. General presentation of flash memory

2.1.1. *The different types of flash memory*

Flash memory is an EEPROM (*Electrically Erasable Programmable Read-Only Memory*). There are two main kinds of flash memory, named according to the logical gate employed to build a basic memory cell: NOR flash and NAND flash. Thanks to the micro-architecture of a NOR flash memory chip, it is possible to address its content at the byte granularity. NOR flash memory shows good performance during read operations: a study [RIC 14a] reports a read latency of about 20 ns. Therefore, this type of memory is used in a context where read performance is vital, i.e. for storage of code dedicated to its direct in-place execution in the flash memory. Writing latency of NOR flash memory is much greater than its reading latency. Usually, NOR flash can be found as a storage support component for operating systems in many simple embedded systems, such as motherboard BIOS code in desktop computers, etc.

On the other hand, the cells of NAND flash memory are addressed by *pages*, i.e. by data packets with a fixed size. With respect to NOR flash, storage density of a NAND flash is greater and its price per bit is lower. NAND flash has a higher read latency than a NOR flash (between 25 and 200 μs according to [GRU 09] and [MIC 06]); however, its write latency is more balanced than in the case of a NOR flash. NAND flash memory is employed in the context of data storage. NAND flash can play the role of secondary storage in complex embedded systems (for example, smartphones and today's tablets), and peripherals for data storage such as SSD disks or USB keys. NAND flash memory can also be found in storage cards in SD or MMC formats or in portable multimedia readers such as mp3 players. As this book is dedicated to storage systems, we will only focus on NAND flash memory.

2.1.2. *Operating physical principles*

The concepts regarding cells of flash memory built from the so-called "floating-gate" transistors are presented below. Another technology exists as well, called "charge trapping" [RIC 14a, TEH 13], but we will concentrate on the first, most common technology.

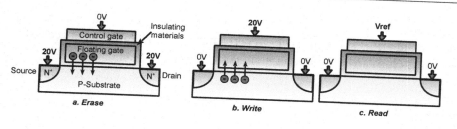

Figure 2.1. *Scheme of a basic cell of a flash memory, composed of a floating-gate transistor, and of the different operations which can be performed upon it: erase (a), write or program (b) and read (c). For a color version of this figure, see www.iste.co.uk/boukhobza/flash.zip*

A basic cell of a flash memory [BEZ 03, FOR 07] is composed of a floating-gate transistor. As illustrated in Figure 2.1, the floating gate is insulated by means of dielectric (insulating) materials. Therefore, electrons trapped in this gate are kept inside for a considerable period of time, providing the flash memory with its property of non-volatility. By applying a high voltage to the control gate or to the substrate, it is possible to charge or discharge the floating gate. Three kinds of operations can be performed on the cell: erase, program (or write) and read, as illustrated in Figure 2.1. The *erase* operation consists of emptying the floating gate of its negative charge, i.e. the electrons contained within. This operation is performed via field electron emission (*Fowler-Nordheim tunneling*). A high voltage, of about 20 volts [FOR 07], is applied to the substrate while a zero voltage is applied to the control gate. *Programming* (writing) consists of negatively charging the floating gate, and it employs field emission as well. In this case, a zero voltage is applied to the substrate while a high voltage is applied to the control gate. Finally, the *read* operation consists of applying a reference voltage to the control gate of the transistor. If the floating gate is charged (negatively), the transistor is turned off and no current is flowing in the channel between drain and source: this situation typically corresponds to a logical "0" (zero) stored in the cell. If the gate is not charged, the transistor is conducting: this is equivalent to a logical "1" (one).

We can distinguish several sub-types of NAND flash cells. *Single Level Cells* (SLC) store 1 bit per memory cell. The term *Multi Level Cell* [FAZ 07] denotes cells which can store 2 or more bits. *Triple Level Cells* store 3 bits. The increase in storage density and the decrease of cost per bit provided by MLC and TLC cells are counterbalanced by a decrease in reliability, endurance (capability of sustaining a more or less significant number of erase operations) and performance. The differences in performance and endurance between SLC and MLC chips are presented in more detail in the following sections.

2.1.3. Simplified hierarchical architecture of a NAND flash memory chip

Figure 2.2 illustrates a simplified architecture of a NAND flash memory chip. This is a high-level vision, relatively abstract with respect to the micro-architectural level. A flash memory chip has a hierarchical structure. It is composed of a certain number of *planes*. Planes contain *blocks*, and blocks contain *pages*. The chip is equipped with an input/output (I/O) bus which makes the reception of commands and addresses possible, as well as the transfer of data to or from the host computer system. This bus is usually multiplexed, which means that commands, addresses and data share the same pins of the flash chip. Depending on the chip version, buses with 8 or 16 bits can be found.

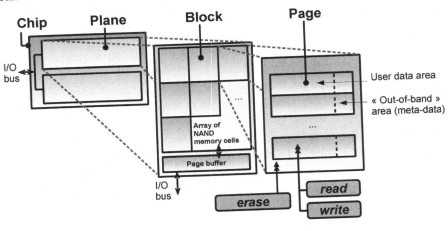

Figure 2.2. *Simplified architecture of a NAND flash memory chip. For a color version of this figure, see www.iste.co.uk/boukhobza/flash.zip*

2.1.3.1. *Page, block and plane*

A flash page (sometimes called sector) represents the granularity of read and write operations on the NAND chip. A flash page is divided into two zones (see Figure 2.2); the first one is called *user data area*. This zone constitutes the largest part of the memory space of a page. It contains the data stored in this chip. The second zone is the *Out Of Band area* (OOB), also called *spare data area* [SAK 07]. It contains meta-data regarding the information stored in the user area. For example, it stores the information required for a correct operation of the management layer of the flash memory: parity information for error-correction codes, meta-data for address translation, bad block markers, etc.

The size of a flash page can vary depending on the chip model. Relatively old chips, (small blocks [MIC 05] models), are composed of pages with 512 bytes of user data and 16 bytes OOBs. Today [1], most pages of SLC chips have 2048 bytes plus 64 bytes of OOB (these versions are called *large blocks*). For MLC chips, the dimensions can be of 2048 (64 OOB), 4096 (128 OOB) or even 8192 (128 OOB) bytes [GRU 09]. In this book, unless stated otherwise, when we refer to the "page size", we refer only to the size of user data. A flash block contains a certain number of flash pages, equal to a power of 2. This number varies, depending on the chip model. Generally, we can find a number of pages per block equal to 32 (small blocks), 64 or 128 (large blocks).

A flash memory chip contains 1, 2 [GRU 09] or 4 [INT 08] planes. The number of blocks per plane varies, depending on the model of the NAND chip. For example, [MIC 05] presents a *small blocks* chip with a total size of 128 MB containing only one plane with 8192 blocks. Another, more recent example, [GRU 09], consists in a 1 GB chip, with one plane containing 4096 blocks. The set of blocks contained in a plane can be called the "NAND transistor array" or simply "NAND array". In addition to this transistor array, a plane contains a register called *page buffer* or *page register* (see Figure 2.2). The size of this register is equal to that of a flash page. It is situated between the NAND array and the I/O bus, and it is used during read and write operations on flash pages (illustrated in Figure 2.3): during a read of a flash page, the page buffer stores the data retrieved from the NAND array. Once that the whole page is present in the register, its content is sent to the host system through the I/O bus. During a write, the page buffer receives the incoming data from the host via the I/O bus. Once this transfer is terminated, the page is sent from the register to the transistor array.

---

1 At the date of writing this book.

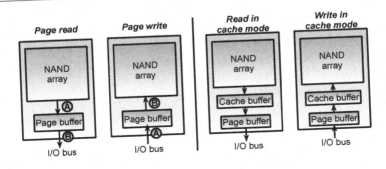

Figure 2.3. *The roles of the page buffer during legacy operations (on the left), and cache mode operations (on the right). For a color version of this figure, see www.iste.co.uk/boukhobza/flash.zip*

2.1.3.2. *The notions of die and Logical Unit (LUN)*

Some studies and technical datasheets of chip models introduce the die notion. A die is a set of planes, generally composed of one or two planes [GRU 09]. A chip can contain one or more dies: two in a *dual die package* model and four in a *quad die package* [JUN 12]. A die is defined as the smallest hierarchical unit capable of executing commands in an independent way within a NAND flash memory chip [MIC 10]. A chip containing several dies is sometimes called a *package*.

It should be noted that according to the ONFI [2] standard [ONF 14], the notions of die and chip are abstracted into a unique concept: the *Logical Unit* (LUN). The LUN is thus defined as a unit that contains a certain number of planes, which eventually share an I/O bus with other LUNs, and that is capable of executing commands independently from other LUNs.

2.1.3.3. *Structure of storage components in an SSD*

Current disks based on flash memory, SSDs (although the only thing they have of a disk is the name, as they are made entirely of electronic components), contain several flash chips (or LUNs). This is due to two main reasons. First of all, the capacity of a flash memory chip is limited, therefore the chips are multiplied within an SSD in order to increase its total capacity.

---

2 The *Open Nand Flash Interface* group is a group of industrial companies (including, among others, Intel, Micron and Sony) that produce NAND chips, whose goal is to standardize the usage methods of the manufactured chips.

Second, the internal architecture of an SSD (i.e. the organization of flash chips and buses which connect them to the controller) with several chips enables the parallelism of certain operations of the SSD, and therefore increases the performance of the storage system.

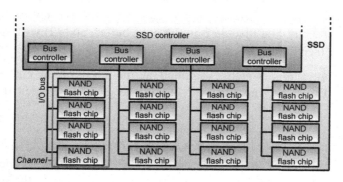

Figure 2.4. *Organization of flash logical units in channels within an SSD. For a color version of this figure, see www.iste.co.uk/boukhobza/flash.zip*

Within an SSD, we can find groups of LUNs [YOO 11b, PAR 09] connected to one another through I/O buses, called *channels*. Each channel is also connected to the SSD controller. This architecture is illustrated in Figure 2.4. The channels have their own controller, therefore it is possible to send two commands to two LUNs situated on two different channels in a completely parallel way. It is also possible to send commands in an "interleaved" mode to LUNs situated on the same channel. The number of channels and the number of LUNs per channel within an SSD are very dependent on the SSD model considered. For example, [YOO 11b] presents some analyses regarding the energy consumption of two SSD models, and the authors report 10 channels and 2 LUNs per channel for the *Intel X25-M* SSD [INT 09] and 8 channels and 8 LUNs per channel for the *Samsung MXP* model [SAM 09].

2.1.4. *Operations on flash memory*

2.1.4.1. *Basic operations*

A NAND chip supports three main operations, called *legacy* operations:

– *Read operation*, which is performed in terms of a page's granularity. A read consists of: (1) reading the page in the NAND array and transferring it into the page buffer of the plane containing this page and (2) the transmission of the page to the host from the page buffer via the I/O bus (see Figure 2.3, page 20);

– *Write operation*, which is also performed in terms of a page's granularity. A write consists of: (1) transmitting the data page to be written from the host through the I/O bus, and placing it in the page buffer; and (2) writing into the NAND array (see Figure 2.3, page 20) ;

– *Erase operation*, which is applied at the level of an entire block's granularity. Therefore, the erase affects all the pages contained in a given block.

The duration of a read operation is, therefore, the sum of the time of transfer from the array to the page buffer and the transmission time through the I/O bus. For a write operation, its execution time is equal to the sum of the transmission time through the I/O bus and the time necessary to program the array. As for the erase, it is a direct operation (without a transmission of data through the I/O bus), although usually a longer one than a read or write. Regarding the execution time and the energy consumption, the values related to these different operations can vary. They depend on the page size, on the number of pages per block, and, more generally, on the NAND chip model concerned. Note that certain chips support the read and/or write of a part of a page ([SAM 07], [MOH 10], [UBI 09a]). In this case, we talk about *sub-page read* and *partial program*.

Some studies have underlined the fact that in the case of MLC chips, we can observe a variability of latency times (and, therefore, of energy consumption) related to read and write operations on flash pages [GRU 09, JUN 12]. These variations are not indicated in the datasheets of the manufacturers of MLC chips. They are due to the micro-architecture of these chips, and as well as the specific programming processes [JUN 12]. Ultimately, the variations of latencies of read and write operations on a page can be relatively large, and the magnitude of these variations depends on the index of the concerned page within the block that contains this page. For example, two NAND chips analyzed in [JUN 12] present variations in read operations within a range between 250 μs and 2200 μs (chip 1), and between 440 μs and 5000 μs (chip 2). SLC chips are not affected by this phenomenon.

2.1.4.2. *Advanced operations*

It is possible to perform the so-called *advanced* operations [JUN 12, HU 11] at the level (A) of a chip and (B) of a structure of chips and channels which can be found in the SSDs. Below we illustrate these operations in a general way.

2.1.4.2.1. Advanced operations at the level of a chip

We can distinguish the following operations:

– Read and write in *cache mode* [MIC 04, MIC 06] are supported in certain chips that contain an additional register of the size of a page, the *cache buffer*, situated between the NAND transistor array and the page buffer (illustrated on the right of Figure 2.3, page 20). This extra register pipelines (temporal overlapping) the transmission of a page to be read from the NAND array into the cache buffer, with a transfer via the I/O bus of a page read beforehand in the NAND array and moved from the cache buffer to the page buffer. The opposite operation is also possible for a write operation;

– *Multi-plane* operations make it possible to send several commands of the same type (read, write or erase) in parallel to different planes of the same chip. This is not a true parallelism, because the planes within a chip share the same I/O bus and therefore data transfers are interleaved;

– The *copy-back* operation moves data stored in a flash page to another flash page situated in the same plane as the source page. This is achieved by means of the page buffer, without any transfer through the I/O bus.

2.1.4.2.2. Advanced operations at the level of a multi-chip and multi-channel structures, such as SSDs

We can distinguish the following operations:

– *Multi-chip* and *die-interleaved* operations consist of sending several cache or multi-plane legacy commands to several chips on the same channel or to several dies of the same chip. By employing the ONFI notation, we can also call these operations *LUN-interleaved* operations. The sub-operations that constitute a LUN-interleaved operation can be of different kinds. They are interleaved because the concerned LUNs share the same kind of I/O bus (the channel); therefore, it is not a parallelism from the point of view of data transfer;

– *Multi-channel* operations consist of sending several commands to several different channels in parallel. In this case, it is a true parallelism.

2.1.4.3. *Latency and power consumption*

The performance and the power consumption of a NAND chip are very dependent on the considered chip model. Table 2.1 illustrates different values for the execution time and the consumption of the three basic flash operations. Note that in the case of read and write operations, these latencies correspond to the transfer from/to the NAND transistor array to/from the page buffer within a plane. In other words, these values do not include the time for the transfer of data via the I/O bus. In [GRU 09], the authors perform some measurements on different chip models, whereas other values are extracted from the datasheets. Certain unavailable values of power consumption are noted with "-". We have already seen that MLC chips can have strong variations in latency, especially during write operations. For this kind of chip, the range of latency values is indicated in the table whenever it is available in the article or in the relative datasheet.

2.2. Constraints and limitations

A certain number of rules have to be respected during the design of a NAND flash memory within a storage system. These limitations are imposed on the one hand by the physical operating principles of basic memory cells, and, on the other hand, by the specific micro-architecture of the NAND flash memory chips. We can classify these limitations into three broad categories: (A) impossibility to perform in-place data updates; (B) wear which gradually affects the memory as the write/erase cycles are performed; (C) unreliability of read and write operations (risk of bit inversion).

2.2.1. *Erase-before-write constraint*

Because of the internal architecture of a NAND chip, it is impossible to write data in a page that already contains data (*in-place* update). Before performing any write operation on a page which already contains data, this page has to be erased. However, the erase operation affects an entire block, and not just a single page. Moreover, this operation has a significant latency. In the literature, this limitation is sometimes referred to under the name of *erase-before-write rule.* This is probably the most important limitation concerning NAND flash memory, because it leads to the implementation of complex management methods in storage systems, which are described in the following sections.

Type	Page size (bytes)	Block size (KB)	Latency (us)			Power consumption (mW)				Source
			Read	Write	Erase	Read	Write	Eras.	Idle	
SLC	4096	256	20	300	1200	19.1	56.0	25.5	13.3	[GRU 09]*
SLC	2048	128	20	200	2000	58.8	78.4	47.6	17.0	[GRU 09]*
SLC	2048	128	20	200	2000	41.1	59.9	35.5	7.1	[GRU 09]*
SLC	2048	128	20	200	500	27.2	35.0	25.3	2.9	[GRU 09]*
SLC	2048	128	20	200	1200	29.9	35.0	20.0	2.9	[GRU 09]*
SLC	2048	128	20	200	2000	35.3	55.2	30.9	2.7	[GRU 09]*
SLC	2048	128	25	200	2000	49.5 - 99			3.3	[LIU 10]
SLC	512	16	12	200	2000	-	-	-	-	[LIM 06]
SLC	2048	256	60	800	1500	49.5 - 99			3.3	[LEE 11]
SLC	2048	128	77	252	1500	-	-	-	-	[KIM 12b]
SLC	4096	512	25	200	700	66 - 165				[LEE 14]
SLC	2048	128	25	200	1500	33 - 165			3.3	[PAR 08]
MLC	4096	1024	30	300-1500	3500	75.9	94.7	70.6	8.5	[GRU 09]*
MLC	4096	512	40	300-1500	3600	66.3	82.3	57.0	11.2	[GRU 09]*
MLC	2048	256	20	300-1100	2800	54.0	58.9	42.4	12.7	[GRU 09]*
MLC	4096	512	110	400-2000	2800	112.0	132.2	111.8	27.3	[GRU 09]*
MLC	4096	512	165.6	905.8	1500	-	-	-	-	[KIM 12b]

Table 2.1. *Latency and power consumption values of basic flash operations retrieved in the literature. *The authors of [GRU 09] have performed measurements on different chip models, whereas the other values are extracted from the datasheets. The authors of [GRU 09] do not cite the brands/models which could contribute to a comparison in terms of performance and consumption. The brands and models of the other chips presented in this table are mentioned in the respective articles*

2.2.2. *Wear constraint*

A flash memory cell can sustain only a limited number of erase operations. After a certain threshold, the cell cannot be used anymore. Given that the erase operation is intimately related to the write operation (see the limitation above), the endurance is calculated in the maximum tolerated number of write/erase cycles. This limitation arises from the fact that the operating window of threshold voltage gradually reduces itself with each write and erase operation on a cell, as more and more electrons are trapped in the insulating layer [RIC 14b]. The limit concerning the number of write/erase cycles depends on the employed NAND flash model. This limitation is typically of 100 000 cycles for a SLC NAND, 10 000 for a two-level MLC and 5 000 for a TLC [BOU 11a].

2.2.3. *Reliability limitation*

A flash memory chip is unreliable in itself. Actually, during read and write operations, phenomena of bit inversion (*bitflips*) may happen in the stored data. These errors are due to the NAND flash memory cells' density inside the chip, and to the high voltages applied to the cells during the write and erase operations [RIC 14b, CHE 07]. Ultimately, this lack of reliability means that without a specific management the data written in the flash memory and that read some time afterwards will not necessarily be read the same. Moreover, in order to reduce the errors due to reliability problems of a NAND flash, during a write operation on the flash it is usually advisable to program the pages within a same block in a sequential way. This is a strong constraint for MLC chips, and a (strong) suggestion for SLC chips [GRU 09].

Data retention is a metric that indicates the time during which a cell is capable of memorizing its content after it was programmed. The retention time of flash memory cells is reported to be between 5 and 10 years [CHE 07]. These numbers should be weighted carefully, as the retention capability of a flash memory cell depends on several parameters, in particular on the number of erase operations performed during the programming. The numbers above suppose a number of erase operations equal to zero. The retention can drop to 1 year for chips at the end of their life [GRU 09].

2.3. Flash memory constraint management systems: general concepts

The existence of constraints in the usage of NAND flash memory requires that every storage system based on flash memory integrates a number of software and hardware components that directly control the flash memory and manage these constraints. We group these components under the term "management layer". This denomination applies to all the storage systems based on flash memory. Two main roles for this management layer can be identified:

1) *Constraints management,* which has been mentioned above, for making the storage system based on NAND flash usable while respecting the rules imposed by the specific technology and micro-architecture of this kind of memory;

2) *Integration of flash memory in computer systems.* Flash memory is a relatively recent technology, and the way it is operated is rather different from the older secondary storage technologies, such as hard disks or optical storage. It is necessary to integrate the storage systems based on flash memory into today's computer systems in the most homogeneous way possible. This is one of the roles of the management layer, which acts as an interface between the flash storage system and the host computer system, while respecting the specific features of this kind of memory.

In conclusion, the role of the management layer is to abstract the specific characteristics and constraints of the flash memory in order to make possible its usage in a computer system. The following paragraphs present the general concepts which are common for any type of flash management layer. For this purpose, we will describe the mechanisms implemented for managing each constraint.

2.3.1. *Management of erase-before-write constraint*

It has been explained above that it is impossible to perform an in-place update of data in a flash memory. However, from the point of view of an application, it is expected to be able to overwrite data in a secondary storage (for example, in the case of an update of a file). The management of this constraint is done at the level of the management system of the flash memory by means of the implementation of a *mapping* process of logical addresses into physical ones. This mechanism works in the following way: the host (application + operating system) addresses the flash storage system via logical addresses. The mapping of the address implemented by the management layer defines the method of calculating a physical address that corresponds to a given logical address. This way, the management layer redirects a data request towards the relative physical position (in the flash memory). This method makes it possible to update data in flash *out-of-place* (physical addresses), in a way which is completely abstract for the operating systems.

2.3.1.1. *Mapping of logical addresses into physical ones*

The example illustrated in Figure 2.5 shows the address mapping principles and demonstrates its usefulness. Let us consider a case where the storage system receives a request to write in a logical page LP_A, which corresponds to a physical page PP_A (step 1 in Figure 2.5). The flash constraints forbid a physical update in-place, i.e. in PP_A. Then, the management layer writes the new version of data (including the update) contained in PP_A into another free physical flash page, for example PP_B (step 2 in Figure 2.5). Once this operation is completed, the management layer updates the address mapping information which indicates that the content of LP_A is now situated in PP_B (step 3 in Figure 2.5). From the point of view of the operating system, an in-place update has been performed.

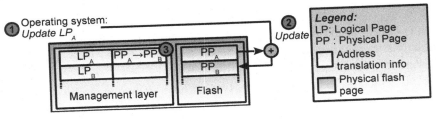

Figure 2.5. *Illustration of the address mapping process. For a color version of this figure, see www.iste.co.uk/boukhobza/flash.zip*

Continuing with this example, the data contained in PP_A is obsolete. For several reasons, it is unlikely that the management layer erases directly the flash block which contains the page PP_A: in fact, first of all, this is a very costly operation in terms of time, and it uses the memory. Moreover, in all likelihood, the block that contains PP_A also contains other pages with valid data that should not be erased. Before being able to erase the block, it would be necessary to move the pages (read and write) containing still valid data, which would be costly in terms of time and wear as well. For this reason, PP_A is marked as *invalid* and the block which contains this page will be erased at a later time. At each instant, a programmed flash page can be either valid or invalid. The flash pages contained in a storage system can be classified in 3 different sets: free pages, valid pages and invalid pages. More generally, we can also talk about an amount of free/valid/invalid space. Note that the notions of valid and invalid states exist only from the point of view of the management layer. From the point of view of the flash memory, there are only free or programmed pages, irrespective of the valid or invalid state of the programmed pages.

As the storage system is used, the amount of valid and invalid data increases, whereas the amount of free space decreases. Therefore, it becomes necessary to recover free space by recycling the invalid space. This role is played by a mechanism implemented in all models of flash management layer: the *garbage collector*.

2.3.1.2. *Garbage collector*

The garbage collector is executed whenever it is necessary to recycle invalid space into free space. The flash management systems usually define a threshold of a minimum free space that, once reached, triggers the start of the garbage collector. Furthermore, some management layers implement garbage collecting processes that are executed in the background: these processes take advantage of the idle time between the I/O requests sent to the storage system to recycle the invalid space into free space. Nevertheless, the presence of a background garbage collector does not exempt the system from a threshold-based garbage collector. In fact, the non-deterministic features of workload applied to storage systems make it impossible to guarantee a certain amount of space recycling in the background.

The garbage collector in the background takes advantage of the idle time between the I/O requests of the workload, and its impact on the response time of these requests is minimal. As for the garbage collector based on a threshold, its execution is usually performed during a write request. When the storage system receives a write request, the management layer verifies whether the threshold of minimal free space is reached. If this is the case, the garbage collector is launched before processing the write request. As explained in the following paragraphs, a garbage collecting operation triggers a certain number of operations (reads/writes/erases) on the flash memory, which result in a considerable execution time. Therefore, this time will affect the response time of the write request that is being processed. In this book, we refer to the garbage collector based on threshold as a *synchronous* garbage collector, and to a background garbage collector as an *asynchronous* one.

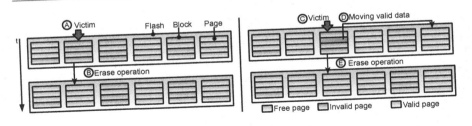

Figure 2.6. *Garbage collector execution in two examples. For a color version of this figure, see www.iste.co.uk/boukhobza/flash.zip*

Although the implementation details of a garbage collector depend strongly on the model of the management layer concerned, we can describe in a simplified way the general execution steps of this process. Two examples of execution of a garbage collector are illustrated in Figure 2.6. In these examples, we consider a hypothetical flash memory containing 6 blocks of 4 pages each. The garbage collector starts by selecting a block to erase, called the *victim block* (A in Figure 2.6). In the case of the example on the left, this block is entirely invalid (all the pages contained in the block are in the invalid state). Therefore, it can be erased directly (B). In the case where the victim block contains a certain number of invalid pages (C), its valid pages contain some user data that must not be lost. Thus, they are copied to a free position (D), and the address mapping information is updated. Once the copying is terminated, the block can be erased (E). Of course, these are very simple examples, and in the real world more complex situations can be encountered.

The time spent by the storage system on performing garbage collection is considered as a self-management time (we talk about *overhead*): it is a time that is not employed for processing I/O requests. In addition, in the case of the synchronous garbage collector, the execution time of a garbage collecting operation delays the execution of the write request which is being processed. Therefore, the performance of the garbage collector is critical.

An important parameter of a garbage collection algorithm is the criterion for choosing the victim block. Garbage collector algorithms of the so-called *greedy* class [WU 94] consider only the number of invalid pages within a considered block, and thus they select as a victim a block that contains a significant number of invalid pages. This is very advantageous for performance, as the number of copying operations of valid data is reduced or even equal to zero. Nevertheless, greedy algorithms do not take into account the wear of the blocks, and their use can potentially lead to a premature wear of the memory. The *cost/benefit* type of algorithms [KAW 95], on the contrary, considers the wear of the memory. These algorithms calculate a score for each block, obtained as a ratio between the estimation of the present wear of the considered block and the number of invalid pages contained within the block. The wear can be estimated from the block's frequency of some systems consider the time since the last invalidation within the block [KWO 07]; others rely on a counter of block erase operations [CHI 99a]. This ratio is thus employed to calculate a score for each block and to find the best candidate for recycling. There also exist other implementations of garbage collectors that depend on the management layer in which they are integrated. Some of these implementation methods will be discussed in the part of this book illustrating the FTL.

2.3.2. *Wear leveling*

Because of the wear limitation of a flash memory, if we seek to maximize the lifetime of the cells we need to optimally distribute the write/erase cycles over the whole memory. In fact, a premature wear of certain blocks can lead to losses or inconsistencies of data stored in a flash memory. Management layers adopt the so-called *wear leveling* policies. These policies are very dependent on the model of management system employed. For example, some of them are integrated in the garbage collector algorithm. For instance, the *cost/benefit* class of garbage collector algorithms presented above [CHI 99a] takes into account the wear of the blocks as a criterion for the selection of the victim block. More generally, the main decisions and methods in which the wear is taken into account are the following:

– *Writing strategy*: when a new logical page of data has to be written, the wear leveling can be taken into account in order to choose the free physical page that will be written;

– *Choice of the victim block by the garbage collector*: here, it concerns the erased blocks. Note that in some systems the wear leveling is separate from the garbage collector;

– *Certain specific methods dedicated to wear leveling*: dedicated methods are implemented in some models of management layers. For example, we can mention the systems which, over time, move the so-called *hot* (frequently accessed) data and *cold* (rarely accessed) data [KIM 12b]. This is done to avoid a large difference in the number of erase operations between the blocks containing rarely accessed data and those with frequently accessed data. On the contrary, cold data, in the absence of a specific management, tends to remain in the blocks that contain them. By regularly inverting the flash positions of hot and cold data, The flash memory management layer limits the difference between the erase operations of its different blocks.

At the management layer level, we can identify the damaged flash memory cells from the wear of the block that contains them: here we talk about *bad blocks*. Despite the wear leveling, bad blocks appear over the life of the flash memory. Some are already present when the memory leaves the factory. The role of the management layer consists in marking the bad blocks and to ensure that these blocks will not be used to store user data [ST 04].

2.3.3. *Reliability management*

From the point of view of the management layer, reliability problems of NAND flash memory are solved primarily by implementing *Error Correcting Codes* (ECC). These correcting codes can be implemented via hardware, by means of dedicated circuits, or via software, for example in the micro-software code executed by the controller on certain peripherals based on flash memory. Generally, *Hamming* types of codes are employed. These codes need to store meta-data in the OOB zone of the pages in the considered flash chip. The small size of the OOB zone limits the number of errors that can be detected and/or corrected by the Hamming codes. In order to increase the robustness of flash-based storage systems, the usage of *Reed Solomon* codes [KAN 09] or *Bose Chaudhuri Hocquenghem* (BCH) codes has been proposed.

2.3.4. *Constraint management systems*

There are two broad categories of flash constraint management systems: *Flash Translation Layer* (FTL) and dedicated *Flash File Systems* (FFS). Both of them implement the principles described above.

FTL is a hardware and software layer, situated in the controller of peripherals based on flash memory, such as SSDs, USB keys, SD/MMC cards and eMMC chips. One of the roles of the FTL is to emulate a block-type peripheral. The concerned peripherals can thus be used the same way as hard disk drives are. FFSs are, on the other hand, mainly employed in the domain of embedded systems. They are used to format embedded flash memory chips soldered directly on the motherboard of systems like in some smartphones, tablets set-top-boxes, etc. The flash constraint management in FFS is performed at the file system level (in other words the operating system), exclusively in software.

Parts 2 and 3 of this book address the FFSs and the FTL, respectively. A more detailed description of these systems is the subject of Chapter 4 for the FFSs and of Chapter 7 for the FTL.

2.4. Conclusion

In this chapter, we addressed the NAND type of flash memory dedicated to data storage. A NAND flash memory chip has an internal hierarchical structure, composed mainly of plane, block and page elements. Three main operations are supported by a chip. The read and write operations target a page, whereas the erase operation concerns an entire block.

The three main constraints concerning flash memory are: (1) the impossibility of updating data within a page without erasing the block that contains it; (2) the wear of memory cells along the repeated write/erase cycles; (3) the unreliability of operations on a flash memory. These constraints are managed by means of: (1) the implementation of a mapping method of logical addresses into physical ones and a garbage collector; (2) wear leveling policies; (3) the usage of error correcting codes. These principles constitute the basis of flash management systems, which can be divided into two broad categories: the FTL and the FFSs.

The following chapter discusses the methods for evaluating the performance of storage systems and, in particular, of flash memory.

3

Evaluation of Performance and Power Consumption of Storage Systems

In this chapter, we will illustrate some general methods for evaluating the performance and power consumption of storage systems based on NAND flash memory. We will consider both the devices based on FTL (*Flash Translation Layer*) and the systems based on FFS (*Flash File Systems*). A performance/power consumption analysis of a storage system is always performed in the conditions of a reference workload, i.e. a benchmark. The first section of this chapter deals with benchmarking for storage systems. The second section introduces the main metrics of performance and power consumption. The third section is a state-of-the-art overview of performance/power consumption analysis studies based on measurements. Finally, the fourth section addresses the simulation-based studies.

The following plan is adopted in this chapter:

1) Flash memory-based storage systems benchmarking.

2) Measurment of performance and power consumption of storage systems based on flash memory.

3) Performance and power consumption metrics of storage systems.

4) Evaluation of performance and power consumption through simulation.

3.1. Benchmarking storage systems based on flash memory

A benchmark is a sequence of reference tests performed on a system in order to measure multiple performance metrics. From the point of view of performance analysis, a benchmark is a reference point that can be useful for evaluating the performance of the system and for comparing the performance of several systems of the same kind, as explained in [JAI 08]. In this section, we describe the state-of-the-art of benchmarks dealing with storage systems. The employed performance metrics (as well as the power consumption metrics) will be illustrated in the next section.

One of the main functions of a benchmark is to produce a workload. The features of this workload are well known and specified in detail in order to guarantee correct comprehension and interpretation of the results. Furthermore, this workload has to be reproducible. In fact, often, an experiment has to be performed many times to confirm or stabilize the results.

In [TRA 08], the authors examine benchmarking methods for storage systems, in particular for file systems. The authors give some general advice. First of all, if a benchmark is used for the purposes of a study, it is crucial to provide a maximum number of details about how the study was performed: which configuration, on which system, with which features, in which environment, etc. This way, others will be able to reproduce the conducted experiments and to validate the results. Second, it is important to justify the choice of the performed benchmark and the related features (configuration, execution environment, etc.). The authors of [TRA 08] review a great number of storage benchmarks used in numerous scientific articles. Benchmarks can be classified into three main categories:

1) *Micro-benchmarks* are tests that perform a small number of operations, generally one or two, on a storage system in a repeated way. Their goal is to study the impact of these specific operations on the system's performance in terms of small granularity, for example, for the purposes of characterization (and, thus, comprehension) and optimization;

2) *Macro-benchmarks* are tests that are composed of multiple operations of different kinds. These tests are intended to represent a real operating workload of a computer system in a particular context of usage. This kind of benchmark is useful to obtain a general idea of performance for a system in production when it is subject to a given workload;

3) *Traces* are sequences of I/O requests that have been registered (traced) on a real system in production, subjected to a given workload. These traces are meant to represent real operating behavior, and they can be replayed during a benchmark test. Benchmarks based on traces are similar to macro-benchmarks, with the difference that the latter are generated synthetically.

In the sections below, a list of benchmarks for storage systems are illustrated. Some of these benchmarks are issued from [TRA 08], which is completed by other benchmarks that are not illustrated in this study.

3.1.1. *Micro-benchmarks*

Micro-benchmarks focus on a small number of operations. They are useful for evaluating the impact of the tested operations on the performance of the studied system, in our case the storage system, with precision. The micro-benchmarks presented here are divided in two categories: those that target a given storage peripheral, independently of the file system, and those that are performed at the operating system's level, which measure the peripheral's performance by taking into account the management software layer for this peripheral. Micro-benchmarks that access a storage device directly are said to be working at *block level* (a block is the granularity of access to a secondary storage peripheral). Another category of micro-benchmarks works at *file system* level. These two types of micro-benchmarks are illustrated in Figure 3.1.

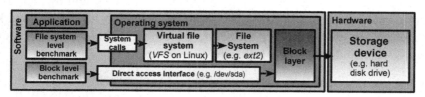

Figure 3.1. *Two types of micro-benchmarks classified according to the level of input of their I/O workload: file system level and block level (the driver is hidden in this diagram for simplicity)*

3.1.1.1. *Block level micro-benchmarks*

These micro-benchmarks perform I/O requests directly on the tested peripheral, i.e. without employing the management software layers (virtual file systems, file system). Most micro-benchmarks directly accessing the storage device target a hard disk drive type of peripheral. For example, we can cite the tools *Open Storage Toolkit* [MES 11], along with *VDBench* [VAN 12]. On Linux, the access to a storage device is done by means of virtual files created in the */dev* directory, which makes it possible to read and write to physical addresses of the peripheral. *uFlip* [BOU 09] is a micro-benchmarking tool that explicitly targets peripherals based on flash memory. Tests are defined on the basis of a sequence of peripheral access patterns. These patterns are, themselves, defined by parameters such as the size of I/O requests, the type of request, the access pattern (sequential/random), etc.

3.1.1.2. *File system level micro-benchmarks*

These micro-benchmarks measure the performance of system calls related to storage. The I/O workload of file system level benchmarks is therefore applied at a high level in the storage system, as opposed to block level benchmarks. In this way in particular, these benchmarks can test the performance of management software layers, especially of the file systems. Numerous file system micro-benchmarks exist. Considering the types of operations tested, we can observe that the vast majority of benchmarks of this kind presented here [BRA 96, COK 09, OPE 13, ROS 92, AXB 14, NOR 03, HEG 08, KUO 02, PAR 90] support, at least, the testing of standard read and write operations on one or several files. On Linux, these correspond to the system calls *read( )* and *write( )*. Simple benchmarks such as *Bonnie* [BRA 96], *Bonnie++* [COK 09], *IOStone* [PAR 90] or *Sprite LFS large/small files benchmarks* [ROS 92] test the performance of a file system by means of creating, reading and writing one or several files. Generally, these benchmarks provide the opportunity to vary the access pattern (sequential/random). More advanced benchmarks add the following functions:

– the possibility to use asynchronous I/O operations is available in benchmarks such as *Aio-stress* [OPE 13], *fio* [AXB 14] or *Iozone* [NOR 03];

– access to files by means of memory mapping functions provided by operating systems is possible in Fio and IOZone. On Linux, this type of access is performed via the call to *mmap( )*;

– certain benchmarks are multi-threaded or multi-process, in order to be able to simulate the behavior of I/O when several processes are executed at the same time in an operating system; we can mention Fio, IOZone, the Flexible File System Benchmark [HEG 08] or TioBench [KUO 02].

While most benchmarks at the file system level focus on performance of the access to data, some of them concentrate on the impact of metadata on the performance of file systems. A benchmark such as *Listrates* [LEN 13] or *Metarates* [LAY 10] can be employed for defining specific test environments which have a great number of files and directories, composing a tree structure with large height and width.

In conclusion, it can be said that micro-benchmarks are used to evaluate the performance of a small number of specific operations, at different levels of the software stack of storage management.

3.1.2. *Macro-benchmarks*

Macro-benchmarks subject the storage system under test to a great number of operations that constitute a complex workload, imitating one of a real application. Therefore, a macro-benchmark is always related to a given context of application, for example, a database, a web server, etc. Thus, macro-benchmarks can be classified according to the targeted application domain. Two main contexts emerge: first, mail servers and file systems in a network; second, databases, especially those related to online transactions.

With regards to the first category, we can cite *Postmark* [KAT 97], which is a benchmark for file systems representing the I/O workload of a mail server. Because of its usage simplicity and its popularity, it is used in many studies, including those related to embedded systems [LIU 10, KIM 12c, LIM 06, KAN 09]. Postmark is a relatively old benchmark and its default configuration is not adapted anymore to the performance of current computer systems [TRA 08]. Therefore, it is necessary to configure this benchmark correctly, in order to employ it at present day. *Filemark* [BRY 02] is an improvement of Postmark that adds *multi-threading* support (Postmark executes only one process) and a smaller granularity of time for the calculation of results. As for the domain of message servers, we can cite JetStress [JOH 13], which reproduces the activity of a *Microsoft Exchange* server. Numerous benchmarks have been defined for testing the performance of file systems in a network. We can cite SPECSFS [SPE 08], NetBench [MEM 01], DBench [TRI 08] or FStress [AND 02]. These benchmarks target file systems in a network such as NFS (Unix/Linux) and CIFS/Samba (Windows).

In the context of databases, in particular those integrated with current web servers, storage systems are subject to potentially very heavy I/O workloads. Furthermore, the storage systems are considered to be a bottleneck for the performance of these computer systems. Therefore, there exists a great number of macro-benchmarks that reproduce the behavior of databases. Some benchmarks, such as *TPC* [TPC 14] and *SPC* [SPC 13], are strongly standardized by committees of experts and they are recognized quite unanimously in the industrial community. These are benchmarks that target storage systems containing databases accessible by means of SQL language. They are presented in the form of specifications, and it is the user's concern to implement them (create the tables, execute SQL requests, etc.). *FileIO* is a benchmark of the *SysBench* suite [KOP 12] that provides a mode called OLTP (*On Line Transaction Processing*), which targets a MySQL database. Further in the database domain, we can also cite *IOBench* [WOL 89]. Generally, the benchmarks presented in this paragraph provide an application load that represents the workload of large scale systems, such as commercial databases or websites with considerable dimensions.

In the embedded domain, a category of benchmarks targeting the performance of Database Management Systems (DBMS), *SQLite* can be found. This DBMS is widely employed in embedded systems because of its implementation simplicity (the database is a single file). *Androbench* [KIM 12c], which targets the Android OS, is an example of this kind of benchmark. Nevertheless, Androbench does not emulate any particular application behavior, and it is more similar to a micro-benchmark.

3.1.2.1. *Other macro-benchmarks for storage systems*

Many studies address the so-called *compilation* benchmarks for evaluating the performance of a file system. Their execution process consists of compiling different applications while measuring the performance of the storage system, for example the total compilation time. Software to be compiled are usually SSH, the Linux kernel, Am-Utils or Emacs. As illustrated by the authors of [TRA 08], the fact that these benchmarks introduce a very heavy CPU load can interfere with the results of tests in which the attention is devoted to the storage system alone. Moreover, it is difficult to compare the results of compilation benchmarks performed on systems with different CPU and/or different compilation toolchains. The *Andrew Benchmark* has been created in order to evaluate exclusively the distributed file system *AndrewFS* [HOW 88].

3.1.3. I/O traces

Trace usage is the third and last category of benchmarks. It consists of (1) collecting traces of access to a source storage system in a real environment; (2) replaying these traces on the same system or another target system while measuring its performance. The traces can be also analyzed for the purposes of I/O behavior characterization of applications executed on the traced system. The replay of a trace is similar to a macro-benchmark in the sense that a recorded trace is meant to represent a real workload: the source system can be traced during its standard execution; in such cases, it is a trace in a real environment. The traced system can also execute a macro-benchmark. Traces can be extracted at different levels of the storage management stack, in a similar way to the two workload injection levels of micro-benchmarks described above: file system level and block level.

3.1.3.1. Frequently used I/O traces in the storage domain

Certain traces are relatively popular and are used in many studies that present new storage systems, in particular to evaluate the performance of these systems. *WebSearch* and *Financial* [UMA 09] are traces available on the website of the University of Massachusetts; they are issued from the tracing of a search engine and from an online financial transaction processing application. The trace *Cello99* [HP 99] contains a one-year-long record of activity of the Cello server working in the laboratory *HP Labs*. Additionally, there are other block level traces available on the website of SNIA (*Storage Networking Industry Association*) [SNI 11]. These traces derive from large scale applications and were registered at block level. They are used, in particular, in studies that present FTL mechanisms (for example, in [GUP 09]).

3.1.3.2. Collection of I/O traces

Traces can be collected by means of an operating systems' code instrumentation at the different levels concerned or by using dedicated tools. At block level, *blktrace* [BRU 07] can trace the access to block-type peripherals on Linux. At system call level, we can mention *FileMon* [RUS 06] on MS Windows, *TraceFS* [ARA 04] and *VFS Interceptor* [WAN 08] on Linux. *ReplayFS* [JOU 05] can replay the traces recorded by TraceFS.

In the embedded domain, the authors of [LEE 12] present *Mobile Storage Analyzer* (MOST), a tool that records traces relative to storage in embedded systems using Android OS (and therefore, the Linux kernel). MOST targets storage systems based on FTL of eMMC type. MOST is built from a modified Linux kernel and the blktrace tool that traces the block level calls to the storage peripheral. Therefore, MOST performs tracing at driver level. Nevertheless, efforts are being made to connect the events traced at driver level to high level information, in particular to the processes that generate the traced I/O operations. On Android, the authors of [KIM 12a] employ blktrace as well for collecting traces at driver level for a storage system on an SD card (FTL). With regard to FFS, these systems do not use the Linux block layer, which excludes the tracers at driver level presented above. As for the tracers at system call level, TraceFS is incompatible with recent versions of Linux kernel (>= 2.6.29). The code of VFS Interceptor is unavailable on the Internet at the date of writing this book.

3.2. Performance and power consumption metrics of storage systems

3.2.1. Performance metrics

There are a great number of performance metrics for storage systems based on flash memories. Two main categories emerge: pure performance metrics (generally related to data transfer speed) and metrics related to wear leveling. In the FFS domain, we can also add the metrics specifically related to the constraints of the embedded systems, and metrics related to file systems in general.

3.2.1.1. Pure performance evaluation metrics

In this context, classical metrics for evaluating storage systems' performance are encountered: at very high levels, the *total execution time of a benchmark* (or, more generally, of a test) is often used to compare different systems or different configurations of the same system with each other. At smaller granularity levels, it is also possible to measure the execution time of a given operation (sometimes called *latency*), for example a read request at block level for a storage peripheral or a *read( )* system call for an FFS. Typically, this latency is calculated as an average value of several calls of this operation, or as a distribution of the execution time of several operation calls for a finer analysis. The *number of flash operations* (read/write/erase) generated by a higher level operation, for example the execution of a *write( )* system call or the execution of an entire benchmark, is also a frequently employed metric. Read and write *throughput* is quite a classical metric as well, and it can be measured at different levels: for example, at application, file system or block level. We can also find metrics that describe the number of satisfied I/O requests per second (IOPS). The IOPS value must always be accompanied by information about, for example, the type of operation concerned, the size of requests, the access patterns, etc. This metric is generally used to describe random access performance with a size of block or request of 4 KB.

Write amplification is a metric that denotes the phenomenon when an application request of writing a certain amount of data causes writing of a greater amount of information in flash. This is due in particular to possible garbage collecting operations that can occur. Note that the existence of caches above the management layer (for example, the Linux page cache situated above the FFS) can cause the system to write less data on the storage device than the amount requested by the application. Therefore, generally, during the calculation of write amplification, it is desirable to consider the workload entering directly into the management layer (independently from the caches). There are several methods for calculating write amplification [HU 09, CHI 11]. An intuitive method consists of dividing the total amount of flash space that is actually written during a test by the amount of space written by the workload which enters the storage system.

3.2.1.2. *Wear leveling evaluation metrics*

Various metrics provide measurements and evaluations of wear leveling. Standard deviation of the counter's distribution of erase operations [LIU 10] of flash blocks is a good indicator.

3.2.1.3. *Evaluation metrics peculiar to FFSs*

Some metrics are directly related to the functions and constraints specific to the FFSs, such as the fact that they are employed in embedded environments. In terms of constraints specific to the embedded domain, the *memory footprint* of a file system is a particularly important metric. Depending on the scenario, the *mount time* can be crucial as well. For file systems that support compression, it is possible to evaluate the *compression efficiency* by comparing the size of the tree structure of files/directories with the physical flash size occupied by a newly created partition that contains this tree. The *impact of compression on performance* can be evaluated by observing the performance of an FFS with and without compression [LIU 10]. Finally, some studies [HOM 09] evaluate the size of metadata that are attributed to an FFS in flash memory, or in other words, the size of flash space which is unavailable for the user.

3.2.2. *Power consumption metrics*

Two main metrics can be found in the literature concerning the energy consumption of storage systems: power and energy. In an electrical circuit, power, whose unit is the watt (W), denotes the power consumption of a circuit at a given instant. The power P of an electrical circuit can be calculated according to the law $P = V * I$, where I is the current and V is the voltage difference between the circuit's terminals. Therefore, it is possible to measure the power in a circuit by connecting a voltmeter in parallel to the circuit, together with an ammeter in series with the circuit's power supply. The power P can also be measured by means of voltage V_{res} across a resistor R inserted along the path of the circuit's power supply, by using the following equation: $P = (V_{circuit} * V_{res})/R$.

Whenever it is necessary to measure the power consumed by a component of a computer system subject to a given workload, it is important to specify the so-called *idle power*. By subtracting this idle power, it is possible to determine the actual power cost due to the activity caused by the workload on the target component.

Energy E, whose unit is the joule (J), is the amount of power consumed over a given period of time: $E(t) = \int_0^t P(t) \, dt$. In practice, the measure of P is discrete and it is always acquired with a certain sampling frequency. Therefore, it is possible to obtain an approximation of the energy with the following equation: $E(t) = \sum_{i=0}^{n} P_i * \Delta t$. The result's precision of the energy calculation is defined by the sampling frequency during the measurements of power.

3.3. Performance and power consumption measurements for flash memory based storage systems

Exploration of performance and power consumption metrics is the first step of the scientific process adopted in order to characterize, model and optimize the values of these metrics in a computer system. Here, we describe the state-of-the-art of different studies and methods dealing with the exploration (1) of performance and (2) of power consumption of storage systems based on flash memory.

3.3.1. *Performance exploration through measurements*

Performance exploration via measurements is achieved by executing a benchmark on a storage system while measuring certain performance metrics. The workload can be applied via a reference benchmark or via a so-called *ad-hoc* benchmark [TRA 08], i.e. which has been developed specifically for the requirements of the study.

At the FFS level, we can find studies specifically dedicated to benchmarking [LIU 10, HOM 09, TOS 09, OPD 10], as well as studies that present new FFSs by comparing them with existing ones by means of a performance study.

The performed measurements can be classified in two broad categories: *time measurements* and *event occurrence measurements*.

3.3.1.1. *Time measurements*

Time measurements belong to the scope of pure performance metrics calculation. Generally, the execution time of an operation is calculated by means of a clock that observes the start and end time of the operation. The execution time of the operation corresponds, thus, to the difference between the end time and the start time:

```
start_time = getTime(clock);
/* execution of the operation */
end_time = getTime(clock);
execution_time = end_time - start_time;
```

On Linux, the clock employed for the calculation is usually the system time. The value of this time is obtained by means of a command or a function like *GetTime*. Among the commands employed to retrieve this time, we can mention the *time* command in Linux. The precision (granularity) of this command depends on the implementations, but usually it is in the order of milliseconds, which is not enough in the case of some fast operations. The Linux system call *time( )* (not to be confused with the command with the same name) presents a granularity of the order of a second. For an efficient measurement of time, the system call *gettimeofday( )* can be used, as it provides a result in microseconds. As for the system call *clock_gettime( )*, it yields a measurement in nanoseconds. Note also that the frequency of a system's time update impacts the time measurements: if this frequency is too low, even a function such as *gettimeofday( )* will not yield a good result; in the case of too frequent calls of *gettimeofday( )* with a clock update at low frequency, measures with the same value of time might be observed. For a finer granularity, for example in order to measure the execution time of a function call in the kernel, it is necessary (1) to use dedicated tracing tools, such as *Ftrace* [ROS 08] or (2) to adapt the code whose execution time has to be measured, as for example in [REA 12]. In this study, the authors indicate legitimately that it is important to minimize the additionnal latencies induced by the instrumentation itself (*overhead*).

When standard benchmarks are being used, time measurements are sometimes performed directly by the benchmark themselves, which employ the same measurement methods as described above. The Postmark benchmark [KAT 97] employs a system call *time( )* at the start and at the end of the benchmark, and it calculates the difference between these two values in order to obtain a total execution time of the benchmark, depending on which other performance metrics are calculated, such as speeds.

3.3.1.2. *Event occurrence measurements and other metrics*

Event occurrence measure is achieved via the usage of trace tools (on Linux we can cite tools such as *SystemTap* [EIG 05]) or the retrieval of statistical values from the storage system. Some peripherals based on FTL support statistical reports by means of SMART (*Self-Monitoring, Analysis and Reporting Technology*) commands [INC 13], which for example can be employed on Linux through the command *smartctl*. In particular, these statistics may concern erase operation counters. As for the FFSs, thanks to their "Open Source" nature, it is possible to modify their code [REA 12, TOS 09, HOM 09].

The memory footprint of an FFS can be estimated by observing the size of the code's segment that is occupied in the object files that represent the compilation result of file system sources [LIU 10], for example via the Linux tool *size*. Furthermore, by means of *free* it is possible to observe the amount of RAM occupied in a system after the mount of an FFS [OPD 10]. As for the extra cost of the metadata in flash memory necessary for the FFS operation, it can be estimated by measuring the maximum size that a file written in a newly created partition (without compression) can occupy [HOM 09].

Wear leveling quality can be evaluated by tracing the occurrences of erase operations and the block indexes in which they are performed [HOM 09].

3.3.2. *Exploration of the power consumption of storage systems based on NAND flash memory through measurements*

Power consumption metrics measurements can be performed for an exploratory purpose, in order to understand and to study the power consumption profiles of the storage system or of one of its components. Besides, power consumption measurements can also be performed during a study in order to validate the precision of a power consumption model, or to measure the efficacy of a proposition of a new storage system or an optimization that targets energy saving. Here we concentrate on the studies that deal with the exploration of storage systems' power consumption based on flash memory as their main subject.

As flash memory is a relatively recent technology, numerous studies illustrate sequences of power consumption measurements that target this kind of memory. Their goal is to characterize the power consumption profiles of these systems. Moreover, some studies further develop this work by analyzing the measurements in order to identify the elements which have a significant impact on power consumption and the elements which, on the contrary, have a negligible impact. Highlighting these elements is a first, essential step in any work for optimizing performance or consumption.

The measurements can be performed at flash chip level [GRU 09, MAT 09]. In this case, a specific hardware platform is required where the flash chip can be inserted and equipped for power consumption measurement. This kind of platform was built by the authors of the two cited studies, and it includes a resistor along the power supply rail of the chip. By means of an oscilloscope, the current at the resistor's terminals is measured. The oscilloscope has to provide a *data logger* function in order to be able to exploit the data at a later time. This method of equipping the power supply line with a measuring device is also employed in many studies that address the power consumption of SSDs [SEO 08, SHI 10, BJØ 10, YOO 11a], and the comparison between the power consumption of SSDs and hard disk drives [DAV 10, LEE 09b, SCH 10a].

3.4. Evaluation of performance and power consumption through simulation

3.4.1. *General concepts*

Simulators are employed whenever the measurements are difficult to perform, because of limitations such as time, cost or simply because of the unavailability of the system concerned. Simulators implement and enact the models that represent the simulated systems subjected to a given workload. Typically, the parameters of the models are set by the user and it is therefore possible to configure the structure, performance, consumption and operating principles of the simulated system. We have already seen that, in the case of a storage system based on flash memory, the management layer defines the access performed in flash for an applied I/O workload. Many simulators take into account this management layer by integrating algorithms that describe the functioning of the specific management layer. These tools target the systems that are based on FTL.

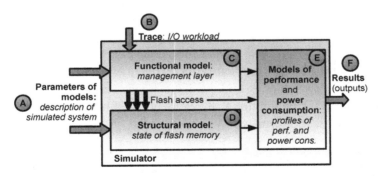

Figure 3.2. *General operation of a simulator for a storage system based on flash memory*

The general operation of a simulator for a storage system based on flash memory is illustrated in Figure 3.2. This diagram is an abstraction of the operation of most simulators presented in this state-of-the-art discussion. The simulators receive a description of the system to be simulated as input (A in Figure 3.2), which corresponds to the parameters of different models implemented in the simulator. Additionally, another input is comprised of a trace (B) that represents the I/O workload applied to the storage system that has to be simulated. This trace is processed by the implementation of a functional model (C) that represents a flash management layer. This management layer calculates the performed flash accesses in order to update the state of the simulated system. This state can refer to the flash state (valid/invalid/free state of pages D in Figure 3.2), and also to the state of other components such as the presence of data in a DRAM buffer in the case of an SSD simulation. Performance and power consumption models (E) are then employed to calculate the metrics values in the simulator's output (F). This process is performed by taking into account the I/O workload, with the possible internal events calculated in correspondence to the management layer and the system's state. The main points that characterize a simulator are the following:

– the *usage context* of the simulator, which refers to general questions such as the research goal during the simulator's operation, the type of simulated system, etc.;

– the *granularity* of models implemented in the simulator: some of them represent a single flash chip, others an entire storage system. As for the performance models, modeling granularity is also an important point. For example, concerning the latencies of flash operations, some simulators model them with a large granularity as constant values [GUP 09] which risks loss of precision in the results; and some model in a very precise way, accurate to a clock cycle [JUN 12] with a risk of increasing the execution time of the simulation because of the system's complexity;

– the different *input* parameters, and the various metrics available in *output*. Most simulators focus on performance, and some simulators offer power consumption metrics as an output. The simulators receive, as input, a description of the system to be simulated and of the workload, and the output usually consists of different statistics concerning the performance and, eventually, the power consumption of the simulated system;

– the *format of the workload received as input*: some simulators require a file representing a trace, and others can generate this workload by themselves;

– the presence or absence of *advanced functions* such as support of multi-chip architecture or advanced flash commands;

– the possibility of *development of new management layers* within the simulator: this can be done if the simulator includes well-defined development interfaces and if the code sections where the user should implement its own management algorithms are sufficiently isolated.

There are several simulators for storage systems based on flash memory [MIC 09b, GUP 09, KIM 09b, HU 11, JUN 12, MTD 08, DAY 13, SU 09, ELM 10, YOO 11b]. Most of them are used for the exploration of design space by varying the different parameters related to the storage architecture and to the simulated management layer. The following paragraphs present the two most important simulators.

3.4.2. *FlashSim*

FlashSim [GUP 09]: two versions of FlashSim exist. The first one, developed in C language, is an extension of the simulator for hard disk drives DiskSim [BUC 08], which is a discrete-event simulator. DiskSim is the most popular tool for hard disk drives based storage systems simulation. Being based on Disksim, FlashSim takes many of its advantages. First, this inheritance allows FlashSim to reuse all the models of satellite mechanisms implemented in DiskSim: buses, caches, queues, etc. Furthermore, thanks to this, it is not necessary to reimplement the fundamental mechanisms of discrete-event simulators, such as time and events management. Finally, DiskSim offers a synthetic generator of traces at block level which can be used for simulating an SSD. There is also another version of FlashSim in C++, separate from DiskSim.

FlashSim, in its C version, was proposed in 2009 in an article that introduced an FTL system called DFTL [GUP 09] (*Demand-based Flash Translation Layer*, described in Chapter 7). FlashSim is employed in numerous studies for two reasons: first, its development infrastructure can isolate in a relatively efficient way the location within the code where an implementation of a management layer algorithm can be done. Moreover, some famous FTL algorithms are already implemented in FlashSim: therefore, it is quite easy for the user to compare its proposition of management layer with these algorithms.

Name	Modeling granularity	Targets the power consumption ?	Complexity of models	Input trace	Complex architecture supported?	Advanced commands supported?	Type	Language	Ref.
NandSim	Chip	No	Simple	NA	No	No	Emulator	C	[MTD 08]
Extension Microsoft for DiskSim	Management + flash (FTL)	No	Average	Block	Yes	Partial	Discrete events	C	[MIC 09b]
FlashSim version C	Management + flash (FTL)	No	Simple	Block	No	No	Discrete events	C	[GUP 09]
FlashSim C++ version	Management + flash (FTL)	No*	Simple	Block	Yes	Partial	Discrete events	C++	[KIM 09b]
SSDSim	Management + flash (FTL)	No*	Average	Block	Yes	Yes	Discrete events	C	[HU 11]
NandFlashSim	Puce	Yes	High	Sequence of flash commands	Yes	Yes	Cycle-accurate	C++	[JUN 12]
EagleTree	Operating system (Management FTL)	No	Average	High level description of applied workload	Yes	Yes	Discrete events	C++	[DAY 13]
[ELM 10]	Operating system	Non	Simple	Amount of data to read/write	NA	NA	Emulator	C	[ELM 10]
FlashDBSim	Management + flash (FTL)	No	Simple	Block	No	No	Discrete events	C++	[SU 09]
[YOO 11b]	Management + flash (FTL)	Yes	Simple	Block	Yes	Yes	Discrete events	C/C++	[YOO 11b]

Table 3.1. Summary of different simulators of performance and power consumption for storage systems based on flash memory described in this state-of-the-art.
*The articles concerning these tools present energy consumption results; however, examining the code of the tools, it is hard to determine how these results have been obtained

3.4.3. *NandFlashSim*

Many simulators accept a block level trace as input as they simulate systems based on FTL, at the granularity of *the flash management layer plus the flash chip* [MIC 09b, GUP 09, KIM 09b, HU 11]. As for *NandFlashSim* [JUN 12], it only simulates the storage's architecture, potentially a complex one, corresponding to a set of chips organized in channels. Therefore, it accepts a sequence of standard and advanced flash commands as input. Thus, NandFlashSim does not offer any model for the management layer. It is distributed in form of a library with well-defined interfaces, intended to be compiled through a program (which is not included with NandFlashSim) that represents the management layer and invokes the library.

Written in C++ language, this library is a *cycle-accurate* simulator. It is possible to describe in detail the latencies in terms of clock cycles for different phases of flash operations, both basic and advanced: command sending, addresses, sending/reception of data, etc. Detailed statistics about the performance and power consumption of the simulated system are available in output after the simulation.

3.4.4. *Other simulators and general comparison*

Table 3.1 illustrates a summary of information regarding the main simulators for storage systems based on flash memories. The comparisons are performed in relation to the elements listed in section 3.4.1.

3.5. Conclusion

In this chapter, the first section addressed the benchmarking storage systems. The main kinds of benchmarks were illustrated, in particular macro-benchmarks and micro-benchmarks. With the former, a system can be evaluated in a general way, and different systems can be compared. The latter are used for a fine analysis of the performance/power consumption of a particular operation. Many benchmarks for storage systems exist, although the embedded domain lacks standardized benchmarks. The second section presented performance and power consumption metrics. Regarding the performance, pure performance metrics and metrics related to flash constraints, in particular to wear leveling, can be encountered. In terms of consumption, the major metrics are power and energy. The third section dealt with the measurement of performance and consumption, classified the performance measurement methods and provided some pointers to interesting state-of-the-art work. The fourth section described the state-of-the-art of simulators of storage systems, employed mainly for evaluating the performance of flash memories. Simulators with different granularity and different goals were portrayed. A comparison between these simulators was performed.

The following chapter describes the file systems dedicated to flash memories.

PART 2

Embedded Domain and File Systems for Flash Memory: *Flash File Systems*

Memory, the warder of the brain
William Shakespeare

Experience is memory of many things
Denis Diderot

4

Flash File Systems

This chapter illustrates one of the large classes of management layers for flash memories: the dedicated *Flash File Systems* (FFS). First of all, a general presentation of these systems is given and we define and explain the three important roles of an FFS: (1) management of storage on a physical device, (2) management of the embedded domain's constraints and (3) organization of stored data and metadata. Next, integration of FFS in present computer systems is described by considering the example of the embedded Linux operating system (OS). Finally, several practical implementations of FFSs are described. The most popular FFSs (JFFS2, YAFFS2 and UBIFS) are presented in detail. Moreover, some information is provided about other FFSs presented in the scientific literature.

The following plan is adopted in this chapter:

1) General presentation of FFSs.

2) Integration of FFS storage systems in computer systems: the Linux example.

3) Presentation of JFFS2, YAFFS2 and UBIFS.

4) Other propositions of FFSs.

4.1. General presentation of FFSs

In the case of file systems for flash memory, all management algorithms are implemented via software at the FFS level, and the latter takes into account the specific features of the employed storage device. FFSs are mainly encountered in embedded systems; they are used in platforms that consist of an embedded motherboard upon which a flash memory chip has been soldered directly. In such case, we talk about a *raw* flash chip or an *embedded* flash chip. The operating system is executed on the processor, and the FFS is integrated within the binary code of the OS. Therefore, the FFSs are a purely software solution. They are used in many embedded systems such as in some smartphones, today's tablets or Set Top Box type of systems (TV decoders, ADSL routers, etc.).

The three main roles of a file system for flash memories are the following:

1) Management of storage and constraints related to flash memory;

2) Compensation of constraints characteristic of the embedded domain, in particular the constraints of limited resources of memory and CPU power, and the resistance to the so-called *unclean* unmounts (accidental unmounts, often caused by a power cut);

3) Traditional management of the file system itself: the way in which data are written and retrieved from the storage device, by means of an organization of files and directories in a tree structure.

4.1.1. *Storage management and flash memory constraints*

An FFS manages the constraints related to the flash memory and implements the concepts illustrated in Chapter 2. The out-of-place updates are performed thanks to an address mapping system. The FFS also implements wear leveling mechanisms. Error correcting codes can be performed by a dedicated circuit or via software within the FFS itself. Wear leveling methods are very dependent on the FFS model.

Like FTL, which associates logical addresses to physical addresses, the FFS associates file blocks (logical point of view) to data packets physically written on the flash memory (see Figure 4.1). The size of these packets can be fixed or variable depending on the FFS. As the physical data packets can be situated in variable positions in flash memory, address mapping is required. In the context of FFSs, we talk about files and data indexing to indicate the association between the logical view and the physical view. In the case of Linux OS, from the logical (OS) point of view, a file is seen as a sequence of adjacent packets of data, called *pages* (not to be confounded with the flash pages). Today, a standard size for a page in a Linux system is of 4 KB.

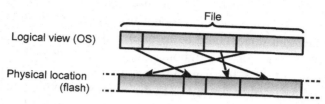

Figure 4.1. *Address mapping, performed by an FFS, associates parts of files (OS logical view) to data packets that are physically written in flash*

4.1.2. *Embedded constraints and scalability of FFSs*

FFSs have to tolerate unclean unmounts. It is necessary to mount a file system before being able to use it, and to unmount it after usage. An unclean unmount corresponds to an arrest of the file system's execution without having launched an unmount command. Typically, this is due to a power cut which causes an abrupt turn-off of the system. In an embedded system, often powered by a battery, this kind of event can happen randomly. An unclean unmount of a file system that is not protected against this event can lead to inconsistencies of data or metadata, and it can cause the loss of all or part of the data. Methods implemented in order to guarantee this tolerance are journaling, usage of structures based on logs and implementation of atomic operations.

FFSs also have to abide by the resource limitation typical in embedded systems (low CPU power, low RAM amount). On the one hand, their RAM usage must be as low as possible, without a significant impact on the performance. In contrast to the FTL, which employs the RAM embedded in the device on which they are implemented, the FFSs use the host RAM to store their internal structures. On the other hand, the algorithms implemented by the FFSs define the workload that these systems bestow on the CPU, and they should demand as little processor resources as possible.

The values of these metrics are, of course, dependent on the implementations of different FFS models. Nevertheless, one problem, particularly highlighted in literature [BIT 05, ENG 05, PAR 06b], can be identified: it concerns the scalability of metrics such as memory consumption (footprint) and mount time of FFSs according to different parameters, in particular the amount of managed flash space (partition size). For some file systems, memory consumption and mount time increase linearly with flash partition size. Mount time should be as brief as possible in devices that are restarted regularly, such as smartphones and tablets. When the increase of these metrics of an FFS is linear with respect to the size of flash memory, its usage becomes impossible, because (1) the memory footprint becomes too large and (2) the mount time becomes too long. Some FFSs present a logarithmic increase; thanks to this behavior, they can manage a much bigger flash space and scale up.

4.2. Integration of FFS storage systems in computer systems: the Linux example

The main roles of a file system are those of (1) ensuring data storage on a device and (2) providing access to data to the user. File systems represent the stored data in form of a tree of files contained in directories. The operating system manages different types of file systems; the OS defines a set of interfaces which an implemented file system has to conform to. The function of these interfaces is to allow the user to manipulate the data stored by the FS, which includes creation and deletion of files and directories, write and read in files, modifications of metadata such as the name or access permissions, etc. The Linux kernel supports three of the most used FFSs, JFFS2 [WOO 01], UBIFS [SCH 09] and YAFFS2 [MAN 10]. Below, we present the principles of integrating an FFS into the stack of software and hardware layers that constitute a storage device based on flash memory in an embedded Linux system.

User space programs (A in Figure 4.2) access files via system calls, such as the primitives *open*, *read*, *write*, etc., sometimes doing this by means of libraries, for example the standard C library. System calls constitute the border between the user space and the kernel space. They are received by an entity called the *Virtual File System*, VFS, (B in the figure), which is detailed in the following section. The I/O requests are transmitted to the FFS by the virtual file system (C). The FFS processes these requests and demands to access the storage device (flash) through a device driver (D). On Linux, all the supported NAND flash chips are accessed by means of a common driver, called the *Memory Technology Device* (MTD), whose details are illustrated below. MTD controls the hardware, i.e. the NAND flash chip (E).

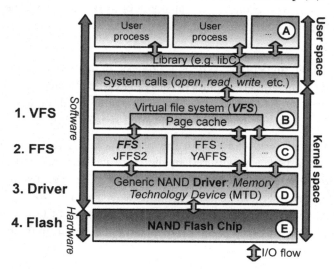

Figure 4.2. *Integration of FFSs into the software and hardware stack constituting a storage system based on embedded flash memory on Linux*

4.2.1. *Linux Virtual File System* VFS

VFS is an abstraction layer of all the file systems supported by Linux; it allows the different types of file systems to coexist inside the operating system during its execution. In fact, thanks to the VFS, the user can access in the same way files that are situated in storage devices or partitions, formatted with different file systems: all read operations are performed via the system call *read*, write operations via *write*, etc. As illustrated above, VFS receives the user's requests and transmits them to the corresponding file system. Besides this abstraction function, the VFS layer includes the following mechanisms of I/O optimization:

1) *Page cache*. It is a data cache in main memory. It buffers all the data read and written in files in order to improve I/O performance. Any data read in a file by a process is read from the page cache, and whenever necessary, it is transferred there from the storage device beforehand. In the same fashion,

any data written by a process in a file is first written in the page cache before being written in the secondary storage. Note that the Linux page cache is a component linked to the memory management system. Nevertheless, in the context of secondary storage management, these two systems are strongly connected;

2) *Read-ahead* algorithm. When a process demands to read a certain amount of data from a file, the Linux kernel can decide to read a greater amount than required, in order to store it in the page cache in anticipation of future accesses. This anticipated read method is called *read-ahead*;

3) Algorithm of *write-back* from the page cache. A data write in a file performed by a process is buffered in the page cache for a certain amount of time. Data are not written directly on the storage device, instead, they are written at a later moment, in an asynchronous way. The *write-back* (or delayed write) improves write performances by taking advantage of temporal locality principle and by assimilating potential repeated updates within the page cache.

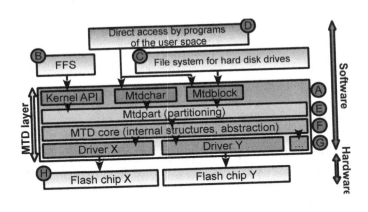

Figure 4.3. *Detailed scheme of the MTD layer, a generic driver for NAND flash chips supported by Linux*

4.2.2. *NAND driver* MTD

The MTD (*Memory Technology Device*) is a software layer of the Linux kernel that centralizes and abstracts the whole set of drivers of NAND flash chips supported by this operating system. Figure 4.3 illustrates the details of the MTD layer. MTD can be viewed as a stack of software layers. The upper layers (point A in the figure) represent the different interfaces of access to the flash device: the access interface of the FFSs (point B, called *Application Programming Interface*, kernel API), and also the character and block devices. These two interfaces (characters and block) are accessible on Linux through virtual devices */dev/mtd* and */dev/mtdblock*. The character device can be employed directly from the user space for read and write operations at physical memory addresses. The block type of devices corresponds to an implementation of a basic FTL algorithm (using block mapping) for a usage by means of a traditional file system (for hard disk drives) (C). In light of weak performances and significant wear caused by the usage of block mapping (see Chapter 7), the employment of this FTL layer is strongly discouraged by the developers of MTD layer [MTD 09]. Thanks to a partitioning layer, MTD offers (E) the possibility of dividing a flash memory chip in several logical volumes, each of which can be formatted separately. The underlying layer (F) contains the structures that are internal to the MTD, and it manages the abstraction of different models of NAND flash chips supported by Linux. Finally, drivers in form of kernel modules, one for each chip model, constitute the lower layer of MTD (G). These drivers manage the models of corresponding chips (H).

4.3. Presentation of the most popular FFSs: JFFS2, YAFFS2 and UBIFS

JFFS2, YAFFS2 and UBIFS are implemented in the Linux operating system and they are used, for example, to store the root file system. They can also be useful for storing data on a separate partition. These FFSs are stable and they are widely employed today[1] in a great number of systems in production. Other propositions of FFSs can be found in the current literature. All the FFSs considered in this section are dedicated to secondary storage of files on raw flash chips. Note that there exist other file systems that manage flash memory as a storage device, which do not correspond to the definition of FFS illustrated here: for example, we can cite the file systems dedicated to be employed in devices based on FTL such as *Flash Friendly File System* (F2FS) [HWA 12, BRO 12].

Let's remark that the term *file system* is used both for designating a model of file system (for example, JFFS2) and for indicating a partition formatted

---

1 At the date of writing this book.

with a given file system (for example, a JFFS2 partition). The Linux operating system views data in a file as a sequence of bytes, from address 0 (first byte of the file) to the address corresponding to the size of the file minus 1 (last byte of the file). This data is divided in adjacent Linux pages. A Linux page has a size of 4 KB in most systems. This logical division of a file in Linux pages is illustrated in Figure 4.4.

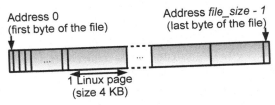

Figure 4.4. *Logical division of data contained in a file into Linux pages, from the point of view of the OS*

4.3.1. *JFFS2*

JFFS2 (*Journaling Flash File System version 2*) [WOO 01] is an FFS dating from 2001 (Linux 2.4.10). Thanks to its maturity, it is still employed in many systems. JFFS2, which supports NAND and NOR chips, is the successor of JFFS, which, however, functions only with NOR type of flash memory.

4.3.1.1. *JFFS2 – general concepts*

With JFFS2, each update to the file system (whether a write in a file, a deletion of a file, a creation of a directory, etc.) is saved to a data and/or metadata packet that is written in flash memory. This packet is called a *node*. Nodes are written sequentially in a block, which is designated as the *current block*. At each instant, there is only one physical block that can be identified as the current block in JFFS2. When this block becomes full, a new free block is selected as the current block. Therefore, a JFFS2 file system can be viewed as a sequence of nodes (a *log*) ordered by writing time, which are physically adjacent in the containing flash blocks. The flash blocks that form the log are not necessarily adjacent within the physical flash memory. This log concept is the basis of the implementation of several FFSs; in fact, it makes it possible to update data out-of-place and, thus, to fulfill the related flash constraint. Furthermore, page writes are sequential within the blocks. The notion of a file system based on a log was first introduced by the file system *Sprite LFS* [ROS 92].

In JFFS2, each node is specifically dedicated to a file. A directory is viewed as a special file without data. There are two main possible types of nodes: (1) the data nodes, which contain data packets, adjacent among each other in the file to which they belong (thus, at the logical level); (2) the so-called *dentry nodes*, which contain metadata about the file (name, access permissions, etc.) and a pointer to the parent directory. The size of dentry nodes is relatively small, of some tens of bytes of metadata, plus the size of the string of characters corresponding to the name of the file. The size of data nodes depends on the size of the write request from the application that triggered the writing of the node in question. Nevertheless, this size is limited to the size of a Linux page, i.e. 4 KB of data plus some tens of bytes of metadata.

As new data updates are performed within the files, new nodes are written and these nodes can completely or partially invalidate the old ones (for example, in the case of data overwriting in a file by a program). Each node contains a version number, thanks to which it is possible to reconstruct the write order of the nodes and, therefore, to determine which node is valid in the case where several nodes contain data related to the same logical positions (*offsets*) in a file.

4.3.1.2. *JFFS2 – indexing and scaling problems*

JFFS2 keeps a mapping table in the main memory, which associates the Linux pages of the files with the nodes that contain the data/metadata related to these files in flash memory. This indexing system is the cause of two main defects of JFFS2. First, this table has to be reconstructed at each mount of the partition. As the nodes have a variable and undefined position in flash memory, JFFS2 has to scan the whole flash partition in question. This is a long operation, whose execution time increases linearly with the partition's size. In a previous study [ENG 05], the authors report a mount time of 15 minutes for a 1 GB partition. Note that the JFFS2 authors propose a function called *JFFS2 summary* [MTD 05], which consists of storing a summary of the nodes that are situated in a block within the block itself. In this way, only the summary information is scanned, which reduces the mount operation's duration. For example, a decrease of mount time from 16 seconds to 0.8 seconds is reported for a 128 MB partition [OPD 10]. The second problem of JFFS2 is the size of the table, which depends on the number of nodes in the file system: the RAM footprint of JFFS2 increases linearly with the size of the file system. Whereas JFFS2 is frequently employed for small-sized partitions (less or equal to 256 MB), due to these properties, it is unsuitable for scaling to greater flash space.

4.3.1.3. *JFFS2 – garbage collector and other functions*

JFFS2 keeps in the main memory a linked list of objects that represent the underlying flash blocks. Each block is contained in a list, which makes it possible to classify it depending on its state. At any instant, a block can be present in only one list, with the exception of the current block. The lists link the blocks according to their state (free/containing valid data/containing invalid data).

The garbage collector of JFFS2 can be launched in two different ways: (1) during a write request, if JFFS2 notices that the free space is critically low, the garbage collector is launched in order to recycle a block; (2) via a Linux kernel thread, the garbage collector takes advantage of timeout between the requests of access to FFS, in order to perform its work in the background. The garbage collector starts by selecting a victim block: typically, it is a block from a list that contains only invalid nodes. It can also be a block from a list with both valid and invalid nodes. Eventually, one time out of a hundred [WOO 01], a victim block is chosen from a list that contains only completely valid nodes. This selection method ensures wear leveling, in order to avoid a significant difference between the erase operation counters of blocks with hot data and those with cold data. When a block with valid data is chosen, JFFS2 copies its data into the current block.

JFFS2 is resistant to unclean unmounts: in the case of sudden power cuts, a scan performed at the following mount is capable of identifying the corrupted nodes (the nodes that were being written when the power cut took place). These nodes are then marked as invalid, and the general state of the file system remains consistent. JFFS2 also proposes an interesting function, which is on-the-fly compression: the data nodes are compressed before being written in flash memory, and they are decompressed before being read by an application. This function reduces the I/O workload of the flash memory, at the cost of increasing the required CPU workload.

4.3.2. *YAFFS2*

YAFFS2 (*Yet Another Flash File System version 2*) [MAN 10, WOO 07] dates from 2002. In particular, YAFFS2 is used in several early versions of

Google's embedded operating system *Android* [YAF 12], which is based on the Linux kernel. YAFFS2 supports only NAND flash chips. This file system is not mainlined in Linux kernel code. Nevertheless, YAFFS2 can be integrated into Linux very simply, by means of a patch provided with the file system sources. Contrarily to JFFS2, YAFFS2 does not support compression.

4.3.2.1. *YAFFS2 - general concepts*

YAFFS2 considers all the entities stored in the file system (for example, files, directories or symbolic links) as *objects*. Data and metadata related to objects are stored in flash memory via data structures called *chunks*. Their size is equal to the size of an underlying flash page, and they are always stored in pages; there is no overlap of one chunk over several pages. Each object has a *header chunk* that contains metadata related to the object in question. For example, the metadata of a file include its name, its parent directory, etc. Furthermore, the objects that represent files also have data chunks, which, as indicated by their name, contain the file's data.

Every chunk has an *object identifier* that links it to the corresponding object. Moreover, each chunk has a chunk identifier, which is an integer, positive or equal to zero. A chunk identifier equal to zero designates a header chunk; a positive integer indicates a data chunk. Furthermore, when it does not equal zero, this identifier indicates the logical position of the chunk's data within the corresponding file. As updates of the file system are performed, the chunks are written sequentially in the pages within a flash block. When this block becomes full, a new empty block is chosen. This mechanism recalls the log structure of JFFS2 file system. An update of data in a file leads to a complete rewrite of the chunk(s) targeted by the update, followed by the invalidation of older update versions. A sequence number is associated to and written with each chunk. This number is an integer that is incremented each time a new block is chosen for being written. The sequence number, similar to the version number in JFFS2, serves to two different purposes:

– Thanks to it, the invalid chunks (containing deleted data/metadata) can be separated from the valid chunks by comparing their sequence number. If two chunks of the same file and with the same identifier value have the same sequence number, because they are within the same physical block, then the chunk that is written at the greatest physical address is the valid chunk. In fact, by definition, chunks are written sequentially in the pages within a block;

– The sequence number is also useful in the case of an unclean unmount, for replaying the chronological write sequence of different chunks of the file system and for re-establishing a consistent state.

4.3.2.2. *YAFFS2 – indexing, garbage collector and wear leveling*

YAFFS2 keeps a table in RAM which maps between the logical addresses in files and chunks in flash. Exactly like in JFFS2, this table is created at mount time via a complete scan of the flash partition. Therefore, the mount time and the memory footprint of YAFFS2 increase linearly with the amount of flash space to be managed and the size of the file system. Nevertheless, YAFFS2 has a function called *checkpointing*, whose role is similar to the summary function of JFFS2. When a YAFFS2 file system is unmounted, the mapping table and several other metadata related to the file system are directly stored in the flash memory in some blocks. These blocks are then marked as containing checkpoint data, and at the following mount YAFFS2 will read only the data contained in these special blocks in order to reconstruct the mapping table in RAM.

The garbage collector of YAFFS2 recycles blocks containing invalid chunks. It is launched on two occasions:

1) After a data update, if the block containing the old versions of invalid chunks is completely invalid (in other words, if it contains only invalid chunks), then it is erased directly;

2) When the amount of free space becomes low, YAFFS2 selects a block with a large amount of invalid chunks, copies the eventual valid data into another position and erases the victim block.

This kind of garbage collector is launched synchronously with I/O requests; therefore, the latency caused by recycling will increase the processing time of the requests. In 2010, a background garbage collector function was introduced in YAFFS2, similar to that described in the case of JFFS2. This mechanism consists of a kernel thread that takes advantage of timeouts between I/O requests to perform garbage collection.

Regarding wear leveling, YAFFS2 does not implement any specific mechanism dedicated to this operation. Nevertheless, the authors of this FFS point out that, by definition, the log structure of this FFS avoids a repeated write on a limited set of blocks [MAN 10]. However, we can argue that the non-separation of hot and cold data, independently from the log structure, can lead to poor wear leveling in specific conditions.

4.3.3. *UBI and UBIFS*

UBIFS [HUN 08, SCH 09] (*Unsorted Block Images File System*) dates from 2008 (Linux 2.6.27). Its implementation addresses certain limitations that are present in JFFS2 and YAFFS2; for this reason, its adoption spread rapidly.

4.3.3.1. *UBIFS – the UBI layer*

In contrast to JFFS2 and YAFFS2, which both directly use the MTD layer to access the flash memory, UBIFS uses an abstraction layer called UBI (*Unsorted Block Images*), presented in [GLE 06b], which manipulates the memory directly through MTD.

UBI is a logical volume manager, which means that it formats and manages the flash directly, while presenting a so-called virtual or logical volume to the upper layers (the FFS, in our case UBIFS). The main function of UBI is to partially manage the flash constraints and to relieve the FFS of some responsibilities related to the management of flash memory, in particular of wear leveling. UBI performs an association between the logical blocks, presented to the FFS, and the physical blocks of the managed flash memory. The logical blocks presented to the FFS have the same size as the underlying flash blocks. However, this association is a simple mapping at a one block granularity level, which means that the FFS still has to perform sequential write operations in logical blocks, and it cannot perform in-place updates therein [UBI 09b]. UBI manages the wear leveling through the following mechanism: when the FFS requests a new empty block to write data to the logical blocks are allocated to this demand. Furthermore, there is a kernel thread related to UBI, which regularly moves cold data, for wear leveling purposes. It is possible to create several partitions for several file systems on a UBI volume, as it could be done, for example, in the case of a hard disk drive. As the wear leveling is done at the level of an entire UBI volume, it is more efficient than when it is performed at the local level of a single partition, as can happen in other FFSs. In fact, the larger the space considered by the wear leveling mechanism, the more efficient its operation.

Therefore, it is possible to distinguish wear leveling as *local* or *global* [HOM 09]. UBI also manages bad blocks, which are marked as such when the number of read and write errors in these blocks becomes too large.

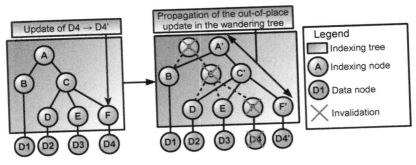

Figure 4.5. *Propagation of an update within the indexing tree of UBIFS. The update of data D4 into D4' triggers the update of the whole indexing branch whose leaf points to D4, up to the root*

4.3.3.2. *UBIFS – general concepts*

Exactly like JFFS2, UBIFS employs the notion of node to designate a packet of data and/or metadata stored in flash. We have already seen in JFFS2 and YAFFS2 that indexing of nodes (or chunks) in a table leads to a linear increase of the mount time and the memory footprint with the size of managed flash space and the size of the file system. To remedy this problem, UBIFS employs a B+ tree [HAV 11]. This tree contains indexing nodes, and its leaves point to data nodes. To avoid a long scan at mount time, the indexing tree is stored in flash; because of this fact, the indexing tree has to be updated whenever a part of data of a file is updated. Regarding the flash memory constraints, it is not possible to delete data in an indexing node, therefore, it is necessary to write a new version of the indexing node (out-of-place update). As a consequence, the parent indexing node, if it exists, also has to be updated out-of-place, and so on, until the root of the indexing tree. The propagation of out-of-place updates in the indexing tree is illustrated in Figure 4.5. As the tree branches move in flash memory, UBIFS authors call this type of tree a *wandering tree*. In practice, UBIFS buffers the updates of the indexing tree in an image of the tree in RAM, in order to avoid flash write operation occuring too frequently. Thanks to the usage of a tree structure for indexing, the UBIFS mount time and memory footprint increase logarithmically with the size of flash space and managed file system. Note that the mount time of UBI increases linearly.

4.3.3.3. *UBIFS – garbage collector and wear leveling*

The garbage collector of UBIFS is launched whenever the ratio of empty flash space becomes low. A victim block is selected, depending, in particular, on the erase operation counters and the ratio of invalid space. Data that is still valid in a victim block are copied to a new position, and the block is then erased.

In contrast to JFFS2 and YAFFS2, UBIFS supports a write-back function for the Linux page cache (presented previously in the section dedicated to VFS), which improves performance, and also makes the system more vulnerable to data losses caused by unclean unmounts. Nevertheless, it is possible to mount a UBIFS file system by deactivating the write-back function (i.e. in a synchronous write mode), by means of *sync* option of the *mount* Linux command. As with JFFS2, UBIFS supports on-the-fly compression of data contained in data nodes.

Table 4.1 presents a comparative summary of the features of different FFSs illustrated here.

Feature	JFFS2	YAFFS2	UBIFS
Types of supported flash memory	NOR, NAND	NAND	NOR, NAND
Type of employed (virtual) device	MTD	MTD	UBI (on MTD)
File indexing structure	Table	Table	Wandering tree
Supported compression algorithms	LZO, Zlib, Rtime	None	LZO, Zlib
Mount time increase	Linear	Linear	Linear (UBI)
Memory footprint increase	Linear	Linear	Logarithmic
Official integration in the Linux kernel	Yes	No (patch)	Yes

Table 4.1. *Comparison of three main file systems dedicated to flash memories supported on Linux*

4.4. Other state-of-the-art FFSs

Scientific literature is rife with propositions of file systems for flash memories. JFFS1 [AXI 99] and the *Microsoft Flash File System* (also called FFS2) [BAR 95] are old file systems for flash memories that date from the

1990s. They are compatible only with NOR flash memory and present several serious limitations. JFFS1 views the whole flash memory as a sequential log in which it writes the nodes that represent the updates of the file system. The main defect of JFFS1 is its garbage collector, which erases the flash blocks in the tail of the log. The data that is still valid are moved to the head of the log. Whereas, in theory, this system performs an almost perfect wear leveling, this behavior should be weighted by considering the strong write workload of moving a potentially great amount of still valid data. The impact on performance is not negligible. As for FFS2, it views the NOR flash as a volume in which each byte can be written only once (remember that NOR flash can be addressed by bytes). The erase operation is done at the level of the whole memory and, thus, it corresponds to a formatting operation. FFS2 employs a chain list to index the files, and, at each update, a part of this list has to be looked at, which leads to low write performances. A linear decrease of performance with respect to the size of the written file is reported [DOU 96].

The mount time is an issue that is addressed by several propositions of FFSs. In order to tackle this problem, a frequently proposed solution is the support of a specific area in flash which contains the information required for a mount operation. This area, of small size, is read at mount time, thus avoiding an entire read of the partition. Some systems employ dedicated blocks [PAR 13, LIM 06, PAR 06b], others use the OOB area of the first page of a subset of blocks, for example ScaleFFS [JUN 08] or MNFS [KIM 09a].

Efforts are also directed towards the file indexing structures, as it is the storage of these structures that determines the memory footprint of the FFS. In order to avoid a linear increase of indexing structures, LogFS [ENG 05] proposes, exactly like UBIFS, to employ a tree, rather than a table. Some systems such as ScaleFFS, propose to store a maximum amount of indexing information in flash while keeping a minimum amount in RAM. MNFS reduces the RAM footprint of indexing structures by using a method similar to the hybrid address mapping techniques employed by FTLs (see Chapter 7). FlashLight [KIM 12c] improves the performances of indexing operations by storing the indexing data in the same pages as the metadata (for example, the name of the file) of the objects that they index.

Certain systems propose an optimization of the cost due to the garbage collector. FlashLight holds information about the valid/invalid state of the managed flash pages. This information exempts FlashLight from reading data contained in the pages of the victim block in order to determine the validity or invalidity of the page, as opposed to systems such as JFFS2. In order to limit the impact of the garbage collector, CFFS separates hot and cold data. Metadata related to files that are written in dedicated blocks; these metadata are considered hot data, since they are updated more frequently then data itself. The latter is thus considered cold data and it is written in distinct blocks. During its execution, the garbage collector mostly selects metadata blocks, because they potentially contain a great amount of invalid pages.

Furthermore, other existing file systems have been proposed to suit the needs of particular and specific contexts. Among these, FlexFFS [LEE 09b, LEE 14] operates on MLC type of flash memory and exploits the MLC capability of being programmed like SLC chips [ROO 08]. When programmed in this way, MLC cells store only one bit; however, they can deliver performance similar to SLCs. FlexFFS partitions the flash space into one SLC area with good performances and a low storage capacity, and one MLC area with lower performance and a greater storage capacity. The partitioning is performed dynamically and it can be adapted to the application's requirements. In the context of information security, DNEFS [REA 12] is an improvement of UBIFS that offers functions for secure data deletion. A secure deletion is achieved when the data deleted by the user are absolutely no longer recoverable from the storage device, even through specialized analysis tools. This function is implemented by DNEFS by encrypting each node with a unique key. During the deletion of data the key is destroyed, which avoids the need to erase data itself and the impact on performance associated with this costly operation.

4.5. Conclusion

In this chapter, we presented the file systems dedicated to flash memories. They represent a solution for managing flash constraints purely via software in embedded systems equipped with raw flash chips. The three major functions of an FFS are the management of flash constraints, the management of embedded constraints and the organization of data storage. The example of Linux kernel illustrated how the FFSs are integrated within the computer systems; on Linux, storage based on FFS can be viewed as a stack composed of the following layers: application, virtual file system, FFS, NAND driver and flash memory chip. Finally, we presented more details of the implementation of three most popular FFSs: JFFS2, YAFFS2 and UBIFS. Other FFS propositions were also briefly described.

5

Methodology for Performance and Power Consumption Exploration of Flash File Systems

Performance and power consumption exploration of a storage system consists of identifying the different components of this system and the parameters that affect performance and power consumption. Exploration is required for analyzing and understanding the results of a study focusing on these metrics. It is also an essential step in modeling the impact of the identified components at a later moment. During the interpretation and the subsequent modeling of these results, it is important to gather a maximum amount of knowledge regarding the operation of the studied system. In such cases, we talk about functional exploration, which has to be performed in addition to experiments for estimating performance and power consumption. This chapter illustrates a methodology applied to FFSs. Several parts of this methodology can be applied to other types of storage management systems. This methodology is issued from a study regarding the modeling of performance and power consumption of systems based on FFS [OLI 16].

This chapter adopts the following structure:

1) General presentation of the exploration methodology.

2) Description of a Linux toolset for estimating performance of systems based on FFS.

3) Description of a tool for estimating power consumption of embedded systems: Open-PEOPLE platform.

5.1. General presentation of exploration methodology

5.1.1. *Methods and tools*

The exploration phase illustrated in this chapter can be described in terms of different sections, represented in Figure 5.1. First of all, it is necessary to understand the operating principles of management algorithms (*functional exploration*), and, second, to identify the storage system's components that affect performance and power consumption of the I/O operations to the storage device. In order to apply this methodology to the FFSs, we perform this exploration at the operating system level. In particular, we focus on Linux system calls *read( )* and *write( )*, which correspond to read and write operations, respectively, on data within files.

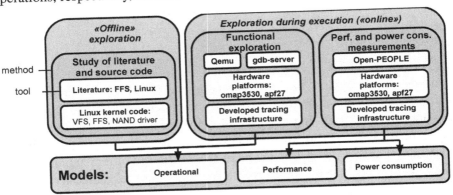

Figure 5.1. *Different complementary methods of exploration and the resulting types of models*

The functional exploration is performed in two parts: first, an *offline* study is performed. In the case of the FFSs, it consists of studying the literature that describes the operation of the management layers in flash storage: books and articles regarding Linux [BOV 05] and the FFSs [WOO 01, MAN 10, HUN 08]. Furthermore, sometimes, a study of the source code (FFS, VFS, NAND driver MTD) can be necessary. In fact, certain implementation details that affect performance and power consumption are not mentioned in the documentation. In this chapter, we focus, in particular, on management by means of the FFS JFFS2 [WOO 01]. The source codes of the FFSs, NAND driver and VFS are integrated in the Linux source code. The study of the Linux kernel source code can be performed in different ways. It is possible to download the Linux code from the official website[1] and to study it by means of a simple text editor. Otherwise, code exploration tools can be used such as *cscope* [STE 85] or online tools such as *Linux cross reference* [GLE 06a].

Functional exploration also has to be performed during the execution of a program (*online*). Several tools can be used for this purpose; on Linux, tracing tools exist at application level (strace, perf, gprof) and kernel level (ftrace, oprofile, systemTap, kprobes). *Gdb-server* [STA 91], together with the *qemu* emulator [BEL 05], is a very useful tool as well. Gdb-server is an implementation of the GNU GDB debugger capable of remotely debugging an application. This tool is frequently employed for system development, in particular for embedded applications in contexts of cross-compiled development with a host computer (a standard PC) and a target execution platform, the embedded system. Gdb-server is not only useful for debugging, but also for exploration: it allows the user to set breakpoints in the kernel code and to monitor the values of different variables during the execution. This is particularly useful in an environment such as VFS, which employs many function pointers. In fact, offline exploration may sometimes not be enough to determine which function is pointed to, whereas this can be done during the execution.

In the case of FFSs, we have developed specific tracing tools that will be illustrated later in this chapter. Concerning the measurements of power consumption, they have been achieved using OPEN-PEOPLE platform for power consumption measurements [SEN 12, BEN 14], which also will be presented in this chapter. Linux version 2.6.37[2] is employed in the studies. The following sections present the embedded electronic boards employed for exploration.

5.1.2. *Hardware platform*

The *Omap3evm* electronic board [MIS 13], a photograph of which is presented in Figure 5.2, is equipped with a *Texas Instruments OMAP3530*

---

1 The code of all the Linux versions is available at https://www.kernel.org/.
2 Many parts of code, as well as the structures, have not evolved significantly since this version.

processor, based on an *ARM Cortex A8* core clocked at 720 MHz. The electronic board also has a *Micron* memory chip with 256 MB of DRAM memory and 256 MB of NAND SLC memory [MIC 09a]. Linux version 2.6.37 (dating from 2011) is executed on the electronic board. In the rest of this document, we will call this board "Omap board". The flash chip integrated in this board contains blocks with 64 pages, each with the size of 2 KB. Most power consumption measures presented here were performed on the Omap board, which has a particular feature: flash memory and RAM memory are situated within the same chip. Thanks to a power consumption measurement point situated along the power rail of this chip, the measures obtained at this point consist of the sum of power consumption of flash memory and that of RAM memory. Another measurement point is situated along the CPU power supply path.

Figure 5.2. *A photograph of* Mistral Omap3evm *embedded board, which represents the main hardware platform used for the experiments related to FFSs that are presented in this book*

5.2. A toolset for performance exploration of FFS-based systems on Linux

Several exploration tools have been developed in the application context of our methodology. These tools are oriented towards the performance of FFS-based storage systems on embedded Linux. They provide measurements of performance and event occurrences for exploration, modeling and benchmarking for this kind of storage system.

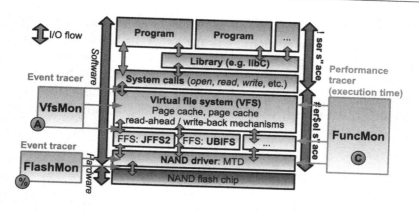

Figure 5.3. *Tracing points related to various developed exploration tools*

These tools help in collecting information about performance and execution of different operations related to I/O requests, and, in particular, about the read and write operations of data in files. The levels of the kernel at which each tool works are illustrated in Figure 5.3. This toolset is composed of three tools:

1) *FlashMon*[3] (*Flash Monitor*) traces information concerning the occurrence of events at the NAND driver level, in other words, at a level close to the hardware [BOU 11b, OLI 14a];

2) *VFSMon* (*VFS Monitor*) traces information concerning various events linked to storage management and to the related parameters of VFS and FFS management layers;

3) *FuncMon* (*Function Execution Time Monitor*) traces the execution time of various operations related to storage management. This is a performance tracer that works at all the levels of the flash storage management software stack.

The tools presented here employ a mechanism, provided by Linux, called *Kprobes* (for *Kernel probes* [KEN 14]). Besides *Kprobes*, there are several tools that perform on-the-fly exploration of events and execution time at kernel level, for example *SystemTap* [EIG 05], *OProfile* [LEV 04] or *ftrace* [ROS 08]. A drawback of SystemTap and OProfile is the need to install software dependencies (libraries and other programs), which can make their

---

3 Available at https://sourceforge.net/projects/flashmon/.

initialization quite cumbersome, in particular in an embedded development environment (cross-compilation). On the other hand, ftrace is natively supported by the kernel; nevertheless, this tool is supported rather poorly on the ARM branch of Linux, in the kernel versions compatible with hardware platforms used in the context of our work. Kprobes is a mechanism natively integrated into the kernel, with simple usage and exhibiting little overhead performance [KEN 14].

5.2.1. *Flashmon*

Flashmon's main goal is to trace the access made to flash memory through the main operations of read and write of flash pages and erase operations of blocks, at the level of NAND driver MTD. The first version of Flashmon dates from 2011 [BOU 11b]. A second version was released in 2013 [OLI 14a].

5.2.1.1. *Flashmon: basic concepts*

Flashmon is a Linux kernel module that traces the function calls corresponding to the execution of basic flash operations at the level of the generic NAND driver MTD. A Linux module is a program containing kernel code that can be dynamically loaded and unloaded during the execution. A Linux module is compiled by means of kernel source code for access to the headers of the latter. Nevertheless, building a module does not require a compilation of the kernel. Flashmon is composed of about 1000 lines of code in C language, and it employs a subtype of *Kprobe* called *Jprobe*. A Jprobe is a probe placed on a kernel function.

5.2.1.1.1. Flashmon's implementation

Flashmon places three *Jprobes* on MTD functions corresponding to read, write and erase operations on flash memory (A and B in Figure 5.4). Each probe is associated with a function, a *handler*, implemented by Flashmon and executed each time a probed function is called. The usage of Jprobes allows the handler to access the values of the probed function parameters, and this allows Flashmon to trace the moment when a flash operation occurs and the address (page or block index) concerned.

The events traced by Flashmon are stored in a buffer in RAM (C in Figure 5.4) in order to minimize the tool's impact: in fact, writing a log of traced events in flash memory on the fly would interfere with I/O operations traced by Flashmon.

Tracing at MTD level makes Flashmon independent of the employed flash chip model, because MTD is a generic driver, independent of the employed FFS type. Actually, the three main FFSs, JFFS2, YAFFS2 and UBIFS, all employ the MTD driver for accessing flash memory.

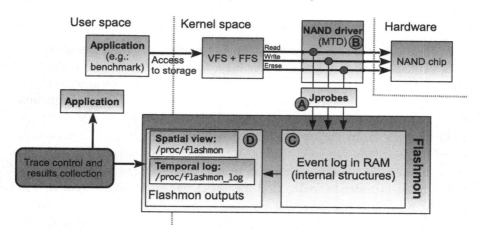

Figure 5.4. *Integration and operation of Flashmon on Linux*

Flashmon traces the following events: read of a flash page, write of a flash page, erase of a flash block and *cache hit* in the MTD read buffer. For each event, the following parameters are traced: (1) address (index of the read or written page, or of the erased block); (2) time at which an event was traced; (3) name of the currently executed process on the CPU at the moment of tracing an event. As will be seen in the following section dedicated to performance exploration, MTD does not support asynchronous I/O operations. This means that during the trace of an event in the MTD layer, the process executed by the CPU at that moment is the process that caused the event. The arrival time of traced operations is measured using the *ktime_get( )* function, which has a granularity of the order of a nanosecond.

Flashmon is released with a script that helps to integrate the module into the source code of an existing kernel and creates a Flashmon entry in the configuration menu of the kernel. Thus, it is possible to compile Flashmon as a standard module, and also to integrate it as a *built-in* function of the kernel. By doing this, Flashmon is loaded at the kernel's start-up, before the mount of the root file system. This approach makes it possible to monitor flash I/O accesses during the kernel boot process. In [OLI 14a], which presents version 2 of Flashmon, flash operations made during the boot process were investigated as a case study.

5.2.1.2. *Flashmon: probed functions*

MTD can be viewed as a stack of software layers. The upper part of this stack contains generic high-level functions for accessing the flash memory, which are known as FFSs. At the bottom of the stack, on the contrary, one can find functions that are very close to the hardware, and whose implementation varies depending on the employed flash chip model (see Figure 4.3 page 62 in Chapter 4). In the case of Flashmon, probing high-level functions is useful because they are very generic. Nevertheless, for several reasons, this can lead to a rather significant error in traced parameters, in particular for the arrival time. In order to understand this issue, we first need to remark that a Jprobe handler is executed just prior to the call of the probed function. In other words, the probed function is executed only at the moment when the corresponding handler returns. Flashmon determines the arrival time of an event within the associated handler using a function such as *getTime( )*. This leads to a potentially significant error when tracing is performed at high level in MTD:

– First, the high-level MTD functions that correspond to read and write operations can trigger a sequence of several flash operations of these types in a single function call. Even if it is possible to determine the addresses of all the reached pages in the handler, the arrival time traced at that moment is quite different from the exact time of occurrence of the operations in hardware. This is particularly true for the last flash operation of the sequence.

– Second, the time traced at the handler level is delayed with respect to the actual time of the operation in hardware: as the traced function is a high-level one, several function calls within MTD are required before the actual execution of the operation in the chip.

Traced functions must be at the lowest possible level, but still being generic enough to be independent from a given flash chip model. At compile time, verifications are performed on the kernel version value in order to identify potential functions to be traced (as the interesting functions are different according to the Linux version considered).

Moreover, when Flashmon is executed (load time for a kernel module is called "insertion"), a search is performed at runtime to finally determine the optimal functions to trace. That search is based on multiple parameters, in particular the flash chip model considered, and the actual FFS used.

5.2.1.3. Flashmon: outputs and trace control

Flashmon yields two outputs in form of virtual files created during the module's insertion, situated in the /proc directory of the root file system (D in Figure 5.4). From the user space, it is possible to read the content of these files for recovering the trace results. The content of these files is created on the fly by Flashmon at each read request, by using trace results stored in RAM. These two files are /proc/flashmon and /proc/flashmon_log.

/proc/flashmon presents an overview of the number of flash operations sustained by each block since the module's launch. This file constitutes a *spatial* view. This file includes a line for each block of the traced flash memory. Each line contains three values representing the number of read and write operations on pages and the number of erase operations received by blocks, respectively. This file is useful for obtaining an instantaneous overview of the amount of operations performed in flash memory up to a given instant. An example of trace with a spatial view is the following:

```
12 2 1
11 2 1
14 2 1
. . .
```

In this trace, each line represents a block of traced flash memory, sorted by physical address. The two first columns represent the number of page read and write operations, respectively, received by a block. The third column represents the number of erase operations received by the block.

`/proc/flashmon_log` represents a log of traced events in temporal order. The following is an example of this log:

```
125468.145741458 ; R ; 542 ; read_prog
125468.145814577 ; R ; 543 ; read_prog
125468.235451454 ; W ; 12 ; write_prog
125468.238185465 ; E ; 45 ; write_prog
...
```

Each line represents a traced event. The first column represents the event's arrival time, expressed in seconds. The second column represents the type of event: R for a read, W for a write, E for an erase operation and C for a cache hit in MTD read buffer. The third column is the event's target address: the index of read and written pages and the index of erased blocks. Finally, the last column represents the name of the process in execution on the CPU during the trace of the event. Moreover, it is possible to write values in the two virtual files in order to control the trace performed by Flashmon. By writing keywords in the virtual files, the user can pause and restart the trace process, or flush the results as well (reset log or spatial view to 0).

At Flashmon's launch, the user can configure the size of the buffer that stores the trace in RAM, which allows the user to control the memory footprint of the module. This buffer is a circular buffer. At load time, it is also possible to specify whether only one flash partition or the full chip should be traced.

5.2.1.4. Flashmon: conclusion

Flashmon proves to be very useful for measuring the workload's impact with regards to the number of performed flash operations, in the context of storage management on embedded flash memory with FFS. Flashmon has been tested on a wide range of Linux versions, from Linux 2.6.29 to Linux 3.8, as well as on several hardware platforms. A study of Flashmon's impact on the traced system has been performed, and it shows that the impact on performance is lower than 6% and the module's RAM footprint is configurable.

5.2.2. VFSMon and FuncMon

5.2.2.1. VFSMon

VFSMon [OLI 14c], just like Flashmon, is a kernel module event tracer targeting VFS level and FFS interface (see Figure 5.3). This tool, which employs Jprobes mechanism as well, is able to collect a lot of information regarding the processing of read and write operations within the VFS layer, at the interface between VFS and FFS (functions *readpage( )*, *write_begin( )* and *write_end( )*). For generic programming purposes, we chose not to gather information internally to the FFS, in order to focus on the generic events that occur independently on the employed type of FFS. The events traced by VFSMon can be classified into three main categories:

1) *VFS inputs* are high-level functions, representing entry points for VFS calls through system primitives. The two main examples are *vfs_read( )* and *vfs_write( )* functions. VFSMon traces the calls to these functions and the associated parameters: identifier of the accessed file, size and address (offset) of data accessed in the file.

2) *VFS internal events* are related to the usage of the cache page and the associated mechanisms (read-ahead, write-back). VFSMon traces cache hits and misses in the page cache during the read operations. Several parameters related to read-ahead are traced; thanks to them, it is possible to monitor the evolution of the size of the prefetching window. No events directly concerning the write-back function are traced, although the behavior caused by this mechanism can be observed indirectly by tracing the asynchronous write calls to FFS.

3) *VFS outputs* correspond to the interfaces between the VFS layer and the FFS, the *readpage( )* function for the read interface, and the *write_begin( )* and *write_end( )* functions for the write interface.

At load time, it is necessary to indicate to VFSMon which partition should be traced. The main advantage of tracing a dedicated partition (different from the root file system) is the possibility of filtering all the operations due to the system itself, which would otherwise disturb the traces if they were performed on the root file system (*rootfs*). Just like Flashmon, VFSMon stores the trace in a circular RAM buffer, whose maximum size is set at load time, and which is allocated entirely at that moment. The advantage of allocating the entire buffer at module's load time is that (1) this makes the RAM footprint constant and (2) it is not necessary to allocate RAM space at each traced event. As the execution time of a Jprobe handler directly affects the performance of the traced system, it is important to perform a minimum amount of processing tasks.

VFSMon creates a virtual file in the */proc* directory. This virtual file can be read for obtaining a trace and written, in order to pause or start again the tracing process or to reset the trace. Below, an example of trace obtained with VFSMon is presented:

```
1   0.20700080;VFS_READ; 1 pages from page 0 - PC MISS
2   0.21297619; SYNC_RA; offset: 0, req_size: 1
3   0.21304926;  ONDEMAND_RA; offset: 0, req_size: 1
4   0.21311542;   RA_SUBMIT; start: 0, size:4,
        async_size:3
5   0.21315772;   RA_FLAGSET; offset: 0, n: 4,
        lookaheadc: 3
6   0.21349157;   FFS_READPAGE; page 0
7   0.22024388;   FFS_READPAGE; page 1
8   0.22620234;   FFS_READPAGE; page 2
9   0.23040542;   FFS_READPAGE; page 3
10  0.23535849;VFS_READ ; 1 pages from page 1 - PC HIT
```

This trace is composed of several lines, one for each event. The first element corresponds to the arrival time, while the second one represents the type of event. Subsequent values vary depending on the type of event. Indentation of an event corresponds to its nesting in the function call stack, corresponding to the VFS code. In this example, a *read( )* call, triggers a *vfs_read( )* call (line 1) accessing the first page of a file. A *page cache miss* occurs. Then, *read-ahead* is called in synchronous mode (line 2). The *read-ahead* call triggers the read of the 4 first pages of the file through calls to FFS (lines 6 to 9). Finally, a new *read( )* call to the second page of the file yields a page cache hit (line 10), because data were loaded previously in this cache.

5.2.2.2. FuncMon

FuncMon [OLI 14c] is a module that traces the execution time of kernel functions. It is not specifically designed for a particular layer of the I/O software stack, and it can be used to measure the execution time of VFS, FFS and MTD functions. Its operation is based on a subtype of *Kprobes* called *Kretprobes*. These probes, once positioned on a kernel function, allow the user to specify two associated handlers. During the system's execution of the probed function, the first handler is executed before the call to the probed function, and the second handler is executed immediately after probed function's return. In order to determine the execution time of the probed function, using *ktime_get( )* function it is possible to obtain the system time before and after the call of this function with a granularity of the order of a nanosecond. The difference between the two values will thus represent the execution time of the probed function.

During the compilation of FuncMon, the user has to specify the name of the kernel functions to be probed. Again, it is also possible to filter the trace by specifying a partition to be probed. Control and recovering of the trace is done by means of an input in */proc* directory. Below, an example of trace performed by FuncMon is presented, with a configuration for measuring the execution time of *jffs2_readpage( )* function:

```
1  0.21343619;0.22009542; JFFS2_readpage ;1751
2  0.22020772;0.22609003; JFFS2_readpage ;1751
3  0.22617234;0.23030926; JFFS2_readpage ;1751
4  0.23037695;0.23439465; JFFS2_readpage ;1751
```

It can be seen that in this example four calls of *jffs2_readpage( )* were traced. The first column represents the call time of the traced function, and the second one represents the moment at which the function returns. The third column contains the name of the traced function, and the fourth one represents the identifier of the process in execution at the moment of the call.

5.2.2.3. *Interoperability of Flashmon, VFSMon and FuncMon tracing tools*

The three modules can be used independently one from another. Conversely, the tracers can be launched in parallel within a kernel. In such cases, although the outputs are separate (virtual) files, it is possible to obtain a single trace containing a merge of the three traces. To do this, it is necessary to merge the traces and to sort the created file according to the first column, which represents the arrival time of the events. This approach is feasible thanks to the fact that the first column of the three traces always contains the arrival time of the event; in the case of all three tracers, this time is issued from the same source, which is the system time, an absolute clock from the point of view of tracers. An example of such a trace is illustrated in Figure 5.5.

5.3. Exploration of power consumption: Open-PEOPLE platform

Open-PEOPLE (*OPEN Power and Energy Optimization PLatform and Estimator* [SEN 12, OPE 12, BEN 14]) is a research project (financed by the

French National Agency for Research, *Agence Nationale de la Recherche*) whose goal was to provide a software and hardware platform for optimization and estimation of power consumption in embedded computer systems. One of the contributions to this project is the *Open-PEOPLE hardware platform*.

VFSMon
Flashmon
FuncMon

```
183202.020700080;VFS_READ; 1 pages starting from page 0 - PC MISS
183202.021297619; SYNC_RA; offset: 0, req_size: 1
183202.021304926;  ONDEMAND_RA; offset: 0, req_size: 1
183202.021311542;  RA_SUBMIT; start: 0, size:4, async_size:3
183202.021315772;   RA_FLAGSET; offset: 0, nr_to_read: 4, lookahead_size: 3
183202.021343619;183202.022009542;jffs2_readpage;1751
183202.021349157;   FFS_READPAGE; page 0
183202.021447157;R;28386;read
183202.021656849;R;28387;read
183202.021828772;R;28388;read
183202.022020772;183202.022609003;jffs2_readpage;1751
183202.022024388;   FFS_READPAGE; page 1
183202.022086388;R;28891;read
183202.022266772;R;28892;read
183202.022438311;R;28893;read
183202.022617234;183202.023030926;jffs2_readpage;1751
183202.022620234;   FFS_READPAGE; page 2
183202.022686388;R;28894;read
183202.022859311;R;28895;read
183202.023037695;183202.023439465;jffs2_readpage;1751
183202.023040542;   FFS_READPAGE; page 3
183202.023104542;R;28896;read
183202.023276619;R;28897;read
183202.023535849;VFS_READ; 1 pages starting from page 1 - PC HIT
```

Figure 5.5. *Trace obtained by merging the outputs of the three tools. In this case, a file read is traced*

This platform is a test bed for measuring power consumption; it contains several development boards on which power consumption of different components can be measured. It is possible to access this platform remotely, via the Internet, in order to perform benchmarks on one of the installed boards and to obtain power consumption measurements at desired measurement points.

The platform contains, among others, an *Omap 3530* board. All the power consumption measurements presented in the following chapter were performed on this board. The acquisition board employed by the platform to perform power consumption measurements is a *National Instruments PXI-4472B* board [NAT 14]. On the Omap board, two measurement points are considered: the power supply line of the CPU and that of the chip containing

RAM and flash memory. The power consumption cost of bus transfers is not taken into account. At the end of a benchmark, the Open-PEOPLE platform sends the results in form of CSV files with power values measured at different measurement points versus time. Unless stated otherwise, sampling frequency of the acquisition board is fixed at 1 KHz (1000 samples per second).

The time range during which measurements are performed corresponds to the entire execution duration of the benchmark; the platform activates the acquisition board just before the start of the benchmark, and deactivates it after an amount of time set by the user: therefore, the execution time of a benchmark should be overestimated. Due to the characteristic constraints of the platform, it is impossible to perform power consumption measurements with precision in a short time interval (shorter than a second). Therefore, it is impossible to obtain directly the power consumption values of very fast operations such as a read of some kilobytes in a file or a flash memory access. In order to obtain this type of value, the following method is employed: the operation, whose power consumption during a benchmark has to be investigated, is launched many times within a benchmark (for example, in a cycle). In this way, an average value can be calculated from the number of performed operations.

Figure 5.6 illustrates two examples of results that can be obtained with Open-PEOPLE platform. The curve on the left represents the power consumed by a component versus time, during a benchmark execution in which a same operation (for example, a *read( )* call) is performed repeatedly many times within a cycle. The following features can be identified in this plot, on the left of the figure:

– idle power consumption before the start of the benchmark (A) and after its end (B);

– the peak of power consumption corresponding to the loading of the benchmark program (C) and its end (D);

– a plateau region of power consumption corresponding to the power consumed during the operation in question. The average power can be calculated from these results; it is represented in this plot by the horizontal line.

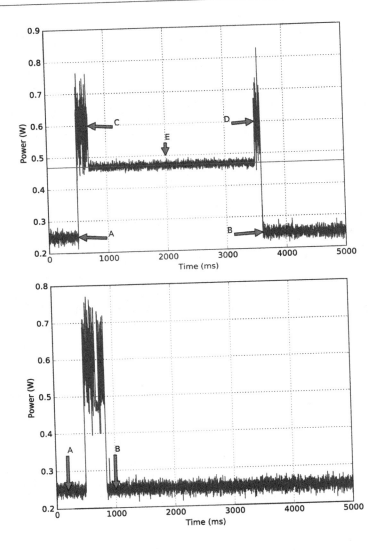

Figure 5.6. *Power consumption measurement on Open-PEOPLE: it is necessary to repeat the operation, whose impact has to be measured, a relatively great number of times (curve on the left). In fact, if the operation is too fast and called an insufficient number of times (curve on the right), the results are difficult to exploit*

If the operation whose impact on performance is measured is (1) too fast and (2) called an insufficient number of times (curve on the right of Figure 5.6), then:

1) Program's loading and unloading time intervals become dominant, and it is difficult to distinguish or to extract the power consumption of the operation in question.

2) The number of samples corresponding to the power of the operation in question is too low for obtaining a meaningful average.

5.4. Conclusion

In this chapter, we illustrated in a general way a methodology for investigating performance and power consumption of dedicated flash file systems. We saw that performance and power consumption measurements sometimes need to be coupled to a functional exploration phase, in order to understand in detail the algorithms of the management layer, and thus to be able to explain the results of the measurement experiments. We also presented some specific tools for measuring the performance of FFSs, as well as the Open-PEOPLE platform for power consumption measurements. These tools are extensively used in the following chapter, which illustrates an application of the exploration methodology presented here.

6

Performance and Power Consumption of Dedicated File Systems: Experimental Results

In this chapter, we will describe the results of the application of the exploration methodology presented in the previous chapter. The goal is to identify all the elements of an FFS-based system that affects its performance and power consumption. For each level of the FFS-based storage management stack on Linux (hardware and driver, FFS, VFS), we perform a number of experiments with performance and power consumption measurements, aiming to highlight the impact of different elements on these metrics. As for the FFS, the case of JFFS2 is analyzed in detail. For every experiment we describe its goal, methodology, results and their analysis.

The outline of this chapter is the following:

1) Study of hardware and driver levels.

2) Study of the FFS level: the JFFS2 example.

3) Study of the VFS level.

6.1. Hardware and driver levels

The hardware components of a storage system define its performance and power consumption. There are numerous such components: first, the type of employed NAND chip determines the performance of basic flash operations and the power consumption during the execution of these operations and during idle state; moreover, the main memory determines the throughput of read and write operations in RAM executed by the management layer, as well as the RAM's power consumption in the idle state and during these transfers; finally, the type of employed CPU determines the execution speed of management algorithms, and, once again, the power consumption related to this component in the active and idle states.

Concerning the performance and power consumption of the main memory, as well as those of the CPU, studying the values of these components represents an important amount of work beyond the scope of this book. Therefore, we consider a simplified view of performance and power consumption of RAM and CPU. Here, these metrics are investigated only in the context of operations related to the FFS-based storage management, i.e. the different functions of the call stack related to processing of I/O requests on Linux. In regard to the NAND flash chip, its performance and power consumption are investigated in detail in this chapter.

6.1.1. *Performance and power consumption of basic flash operations*

Goal: To observe the evolution of performance related to flash memory access at the driver level, depending on the performed operation (read, write, erase), on the number of operations performed during a test and on the access mode (sequential or random).

Method: Micro-benchmarks at the driver level are performed. They are made using Linux kernel modules that employ kernel internal MTD data structures in order to directly access flash memory, i.e. the physical addresses. Three modules were developed, one for each type of operation. Each module receives a parameter corresponding to the number of operations to be performed. At insertion, the module launches the operations in a loop, whose execution time is measured using *ktime_get( )* primitive, with a granularity of the order of a nanosecond. An example of pseudocode for a micro-benchmark of a read operation is the following:

```
------------------------------------
input: n, number of pages to be read
output: t, total execution time
------------------------------------
start = get_time()
for (i = 0 ; i<n ; i++) do:
| read flash page at index i
stop = get_time()
t = stop - start
return t
```

Note that time measurements are directly integrated into the benchmark. As the tests are very simple, the developed measurement tools presented previously have not been used for this experiment. The tests were launched on Omap board (Linux 2.6.37) which is equipped with a flash chip containing pages of 2 KB and blocks of 64 pages. The number of pages to be read and written varies from 2 to 32768 according to powers of 2. The number of blocks to be erased varies from 2 to 512. Before each write test, flash memory was erased. The tests are performed on a partition of 100 MB.

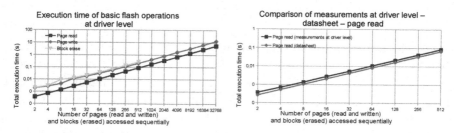

Figure 6.1. *On the left: execution time at the driver level of loops performing a given number of basic flash operations of the same kind. On the right: comparison of measured execution times with values issued from datasheets*

Results: the results are illustrated on the left-hand side of Figure 6.1. They clearly show that the execution time of each operation increases linearly with the number of accessed pages and blocks. Note that in this case, the I/O accesses are performed sequentially. With random accesses, an almost negligible variation is observed: between 5 and 10 μs per flash operation [OLI 13]. By comparing these values with those provided by the datasheet of the flash chip [MIC 09a], it is noticed that the response time is slightly greater for measurements performed at the driver level. On the right-hand side of Figure 6.1, this difference in the case of read operations is illustrated. Similar observations can be done for write and erase operations. Presuming that information provided by the datasheet is precise, the time difference represents an overhead caused by the driver. This overhead corresponds to the CPU time and RAM transfers allocated to the management of MTD driver.

In regard to the power consumption of these operations, the following experiment was performed:

Goal: Monitoring power consumption during flash access at the driver level.

Method: Power consumption is measured along the power supply line of the CPU and the memory chip as a function of time, during the read of 32768 pages, write of 32768 pages and erase of 512 blocks. These values are chosen so that the benchmarks have a sufficiently long execution time, in order to obtain interpretable results, as the access to a single page or block is almost invisible on the measured power consumption curve. Power consumption is also measured in idle state. Let's recall that RAM and flash memory share the same power supply path on Omap board.

Results: Figure 6.2 illustrates the evolution of measured power consumption. First, intervals of power values, relatively stable during the flash access loop, can be distinctly identified. The peaks of power at the beginning and at the end of test correspond to the loading and the shutdown of the test program. They bear no relation to the actual flash memory access. The overhead with respect to idle power consumption on the memory chip is rather low: approximately 0.025 W (15% of idle power consumption on memory chip) for the write and erase operations, and 0.017 W (11%) for the read. At CPU level, the cost is more significant: approximately 0.15 W (56% of idle power consumption on CPU) for the write and erase operations, and 0.2 W (74%) for the read.

6.1.2. *Impact of the MTD read buffer*

MTD implements a buffer in RAM, which is used for read operations. This buffer has the size of an underlying flash page. Each time a flash page is read, data from this page are stored in the buffer. A future read access to this page will then be satisfied from the buffer, without triggering an access to flash memory. A flash page remains in the buffer as long as it is not erased or written in flash memory.

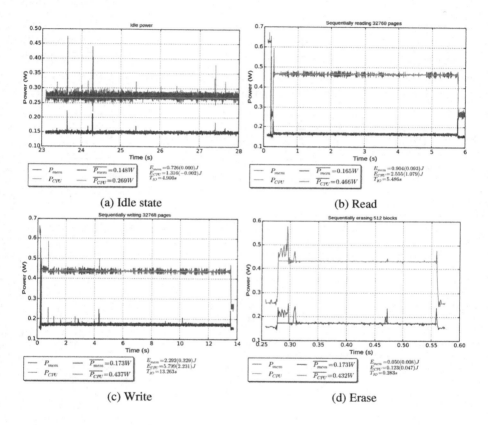

Figure 6.2. *Power consumption observed on the memory chip (flash memory+RAM) and CPU of the Omap board in idle state (a), and during the application of basic flash operations at the driver level: read of a page (b), write of a page (c) and erase of a block (d). For a color version of this figure, see www.iste.co.uk/boukhobza/flash.zip*

Goal: Evaluating the impact of MTD read buffer on performance.

Method: A modified version of the previously used read micro-benchmark; this version accesses more than 32000 times the same page in flash memory.

The results yielded by this benchmark are compared with those from the benchmark concerning sequential read operations.

Results: As for performance, the results are presented in Table 6.1. We can see that accessing the same page many times is a much faster operation than accessing many different pages. In the case of Omap board, the whole amount of accessed pages corresponds to 64 MB of data (pages of 2 KB). During the experiment that employs extensively the read buffer, much higher throughput can be observed than that of a flash memory. The trace of flash operations performed using Flashmon confirms that the flash page is read only once.

Operation (driver level)	Execution time (s)	Throughput (MB/s)
Read 32768 different flash pages	5.49	11.6
Read the same page 32768 times	0.16	400.0

Table 6.1. *Impact of MTD read buffer on performance*

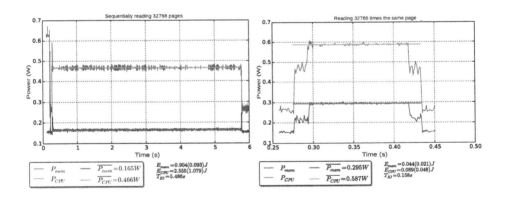

Figure 6.3. *Impact of MTD read buffer on power consumption. For a color version of this figure, see www.iste.co.uk/boukhobza/flash.zip*

Regarding power consumption, Figure 6.3 presents the evolution of the power consumed as a function of time for the two experiments, on the memory chip and the CPU of the Omap board. Again, a strong performance improvement is observed, and it can also be seen that the power on CPU and memory chip is higher during the usage of the buffer. An increase of 0.13 W in the memory chip (88% of idle power of this chip) and an increase of 0.12 W in the CPU (45% of idle power) are observed.

6.1.3. *Conclusion*

The power consumption and performance of the storage system thus depends on:

– hardware features: latency of flash operations, average power observed during each operation on flash memory, RAM and CPU;

– number of performed operations in relation to a given workload;

– overhead in terms of time and power consumption, due to the driver itself;

– positions accessed by read operations of the I/O workload, which will make a more or less intense usage of the read buffer.

6.2. Exploration at the FFS level: focus on JFFS2

The NAND driver is used by the FFS to perform flash I/O operations. In this section, we explore the impact of the FFS part of the management layer on performance and power consumption. As underlined previously, we focus in particular on data read and write operations in files. Let's consider the example of the JFFS2 FFS. In case of a Linux page read, VFS calls *jffs2_readpage( )* function in the case of JFF2. Regarding a write in a page, two functions are called consecutively: *jffs2_write_begin( )* and *jffs2_write_end( )*. These three functions constitute the interface between VFS and JFFS2 file system. At MTD level, a performance and power consumption overhead has been observed, due to the driver, with respect to latency and power consumption values at hardware level (flash chip). The FFS adds an overhead with respect to the driver level as well.

6.2.1. *Read at the FFS level*

6.2.1.1. *General concepts*

Goal: Investigating read performance of JFFS2.

Method: For this purpose, a first experiment consists of measuring the execution time of *jffs2_readpage()* function, while performing simple accesses (sequential and random read in a file). The following experiment is performed: a file with a size of 400 KB is created within JFFS2 partition. The size of 400 KB corresponds to 100 Linux pages, thanks to which it is possible to perform a relatively high number of read operations on different Linux pages during the experiment. The file is created on a newly erased JFFS2 partition, in order to avoid any interference from garbage collection. The file is created and written sequentially with random data using *write( )* (Linux), page by page. This file is then read sequentially, page by page, by means of a user space program performing *read( )* calls in a loop. The size of a *read( )* request does not matter much, because it is reduced to a Linux page size by VFS before the call to FFS. In most experiments presented here, the size of request is thus fixed at 4 KB. The Linux page cache that buffers the access to files is emptied before the test's launch in order to avoid its impact. Using Flashmon and Funcmon, the execution time of *jffs2_readpage( )* function is measured, along with the number of flash pages read during each call of this function. Furthermore, the indexes of physical flash pages being read are traced. The I/O operations on the file are performed (A) sequentially and (B) randomly.

Results: The results are presented in Figure 6.4. At the top, the results of sequential accesses are illustrated. Regarding the execution time of *jffs2_readpage( )*, it can be seen that most calls take about 400 μs. Some peaks are observed, with a value of about 550 μs, and others approximately 600 μs. By studying the number of flash pages read during each *jffs2_ readpage( )* call, we observe that most calls to this function trigger a read of two flash pages. However, some calls trigger a read of three flash pages: these events correspond to peaks at 600 μs. By looking at the indexes of physical flash pages that were read, we can remark that JFFS2 performs sequential read operations. Four block changes are noticeable (changes of read pages' indexes in the bottom curve of Figure 6.4(a)), showing that the file is distributed over five flash blocks. This means that the JFFS2 nodes composing the read file are distributed sequentially in flash memory, which is a normal situation as the file was written sequentially on a newly created partition.

Further considering the case of sequential access, and in particular considering the peaks of 600 μs, it can be observed that these peaks are periodic. The explanation to this periodicity is the following: as the file is written (Linux) page by page, most JFFS2 nodes composing the file contain data that correspond to a whole Linux page of this file. The file's Linux page 0 is contained in a node, page 1 in another node, etc. Therefore, in our experiment, *jffs2_readpage( )* reads only one node at each call. These nodes are distributed sequentially in the flash blocks that contain them. The size of these nodes is roughly of 4 KB: they contain 4 KB of data plus some tens of bytes of metadata constituting the header of JFFS2 node. Therefore, the distribution of nodes in flash memory corresponds to the diagram represented in Figure 6.5.

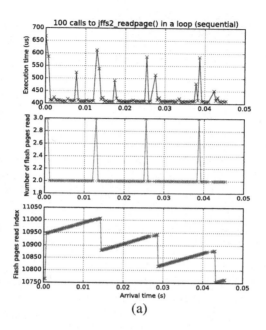

(a)

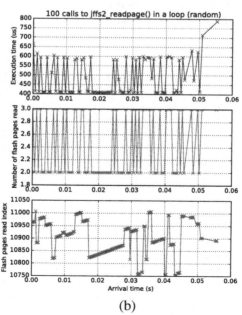

(b)

Figure 6.4. *Exploration of* jffs2_readpage() *performance: (a) the sequential accesses are presented; (b) the random ones. Concerning the three curves in each figure, at the top, we can see the execution time of each* jffs2_readpage() *call as a function of its arrival time; at the center, the number of flash memory pages read by each*

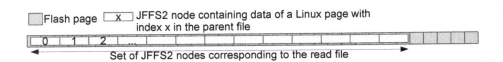

Figure 6.5. *Flash memory distribution of JFFS nodes of a file written sequentially (Linux) page by page in a free partition*

Therefore, with such a distribution, a node is usually distributed over three flash pages. The sequential read of nodes triggered by a sequential read of the file requested by the application will thus prompt a read, sequential as well, of flash pages (this situation is observed on the bottom curve of Figure 6.4(a)). The read of the last flash page containing a node prompts a transfer of this flash page to MTD read buffer. The read operation regarding the following node will then start with a cache hit in this buffer, because the nodes are positioned sequentially in flash memory. Therefore, a read by means of *jffs2_readpage( )* typically triggers $3 - 1 = 2$ reads of flash pages (one node distributed over 3 pages to read, minus one cache hit). Nevertheless, sometimes some nodes can be aligned at the beginning of a page: in this case, no advantage is provided by the MTD read buffer and 3 flash pages are read. This behavior is illustrated in Figure 6.6, and it is verified with the help of Figure 6.4(a) (middle curve).

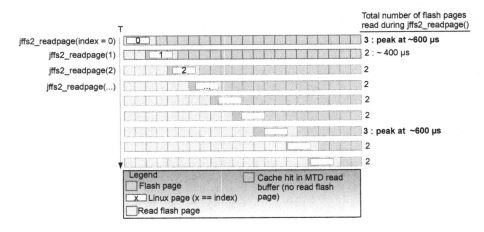

Figure 6.6. *Sequential read of a set of JFFS2 nodes distributed as defined in Figure 6.5*

As for the random I/O accesses, the results are presented in Figure 6.4(b). It can be observed that the number of peaks approximately 600 μs is much greater than previously. In fact, the number of calls to *jffs2_readpage( )*, each of which triggers 3 reads of flash pages, is much greater. This is due to the fact that the behavior observed in the case of sequential accesses is not applicable here: as the I/O accesses at flash level (see Figure 6.4(b)) are less sequential, the MTD read buffer is much less used. Total execution time is also much greater in the random case: about 0.055 seconds against approximately 0.045 seconds of the sequential case.

Regarding the peaks around 500 μs, present both in the sequential and random cases, it should be remarked that their impact is negligible. In Figure 6.7, the distributions of the execution time of *jffs2_readpage( )* for the two experiments (sequential and random) are shown. We can see that the number of peaks at 500 μs represents a negligible percentage (lower than 5%) of the values of execution time.

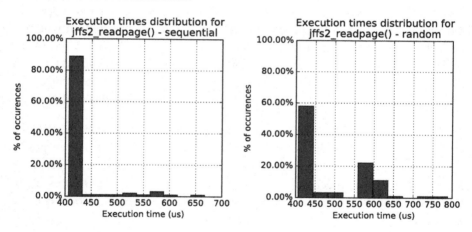

Figure 6.7. *Distribution of execution times of* jffs2_readpage( ) *in the sequential (left) and random (right) cases*

After having studied performance, we will now focus on power consumption by describing the following experiment:

Goal: Observing power consumption during read accesses to JFFS2.

Method: Power is measured on the Omap board during a sequential and a random read of 3000 Linux pages within a JFFS2 file. The rather large number 3000 is chosen in order to be able to perform a sufficiently long test and, therefore, to obtain an exploitable power consumption curve.

Results: The results are presented in Figure 6.8. In addition to the differences regarding the execution time, regular power consumption peaks can be observed in the sequential experiment, which are absent in the random one. These peaks correspond to frequent *RAM accesses*. In fact, in the sequential case, read-ahead mechanism quickly reaches its permanent regime. In our experiments, the maximum amount of data that *read-ahead* can prefetch is of 32 Linux pages (128 KB). Therefore, packets of 32 Linux pages are regularly prefetched from flash memory (execution time between two peaks), followed by 32 cache hits in the page cache. When a process reads a Linux page from the page cache, a copy of this page is created in RAM. The page stored in the page cache is situated in a memory space reserved for the kernel and not accessible by a process executed in the user space. This page is thus copied to the space addressed by the process (user space), into the user buffer, which is passed as an argument of the system call *read( )* at the moment of its invocation. This task is carried out at the VFS level thanks to *copy_to_user( )* function. Therefore, the observed power consumption peaks are caused by sequences of 32 memory copying operations, very closely spaced in time.

The reason why the memory copying operations are not visible on the curve is the low sampling frequency of the system that measures power consumption. In fact, although the number of cache hits is lower, due to the fact that read-ahead mechanism is less efficient with random I/O accesses, in this experiment there are as much memory copying operations between the kernel space and the user space as in the sequential experiment. Nevertheless, the involved copies concern a small number of pages and they are less frequent; they are fast operations, therefore they are not visible. It is possible to observe them by increasing the sampling frequency to 100 KHz. This inspection is presented in Figure 6.9, regarding the experiment with random accesses. A high sampling frequency makes the curves less legible and hardly suitable for processing, because, obviously, the number of samples increases with frequency. Therefore, most results presented here were acquired at 1 KHz.

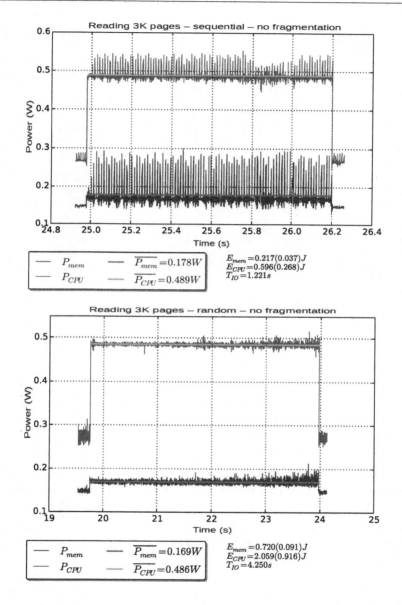

Figure 6.8. *Evolution of power as a function of time during sequential reads (top) and random reads (bottom) of JFFS2 files. For a color version of this figure, see www.iste.co.uk/boukhobza/flash.zip*

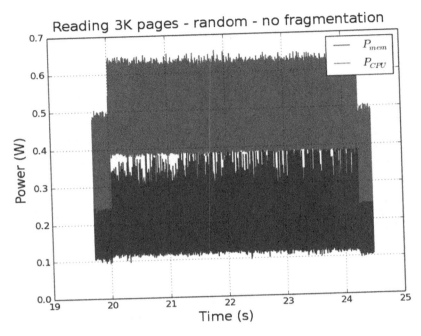

Figure 6.9. *Random read of 3000 Linux pages in a file, with power consumption measurements performed at a frequency of 100 kHz*

6.2.1.2. *Impact of fragmentation*

In previous experiments, the read file was created in a free partition, and it was written sequentially, page by page. This yields a very ordered distribution in flash memory: the JFFS2 nodes contain whole Linux pages and they are distributed sequentially within the blocks that contain them. However, this is an ideal situation. In a situation closer to reality, a random write access to a file yields a fragmentation of the corresponding nodes in flash memory. The following experiment is performed:

Goal: Investigating the impact on the performance of a file's distribution in flash memory.

Method: Two files of 250 KB are created in a new JFFS2 partition built after an erase. The first file is not fragmented; it is created by a sequential writing of data. The second file is fragmented[1]; it is created as the first one and then, 500 packets of data, of 512 bytes each, are written at random positions in this file. The effect of this process consists in creating a very large fragmentation of JFFS2 nodes in flash memory, as illustrated in Figure 6.10.

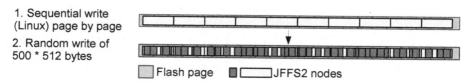

Figure 6.10. *Creation of a fragmented JFFS2 file: at first, the file is created sequentially, page by page, then, it is updated at random offsets with small packets of data.*

Then, each file is read (Linux) page by page and the difference in performance is observed.

Results: The results are presented in Figure 6.11. This figure illustrates the execution time of each *jffs2_readpage( )* call plotted as a function of its arrival time (x-axis). It can be clearly seen that total execution time is much longer in the case of reading the fragmented file: 0.16 seconds against 0.04 seconds for the non-fragmented file. Furthermore, the execution time of *jffs2_readpage( )* functions is observed to be mostly long and very variable for the fragmented file. This can be explained by the fact that there are a great and variable number of nodes composing each Linux page to be read in the case of the fragmented file. Regarding power consumption during the read of a fragmented file, a curve, quite similar to that of the non-fragmented file, is observed (see Figure 6.8), although the duration of the operation is, of course, more significant.

To conclude the analysis of read operations at the FFS level, in the case of JFFS2, the main factor that affects performance and power consumption is the number of pages read by the *jffs2_readpage( )* call. This number depends on the access mode (sequential or random) which makes a more or less intense usage of the MTD read buffer, and on the fragmentation of the file in nodes within the flash pages.

---

1 Note that this is not the usual so-called (external) fragmentation as random reads on flash memory are generally as good as sequential ones.

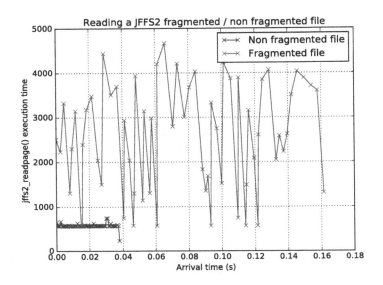

Figure 6.11. *Sequential read of a non-fragmented and a fragmented file*

6.2.2. *Write at the FFS level*

6.2.2.1. *General concepts*

jffs2_write_begin( ) and *jffs2_write_end( )* are the two functions called during a data write operation within a Linux page of a file. The following experiment is performed:

Goal: Investigating the impact of the request size on performance of these two functions.

Method: We know that the write buffer of JFFS2 can buffer the write operations with a size smaller than the size of a flash page. Therefore, the following experiment is performed: thanks to the developed tracing infrastructure, the execution times of these two functions are traced, together with the number of flash pages written during the execution of *jffs2_write_end( )* (*jffs2_write_begin( )* does not trigger any write operation in flash memory). The following benchmarks are launched:

1) Sequential write of 100 Linux pages in a newly created file. Thus, at the end of the experiment, the file's size is equal to $100 * 4 = 400$ KB.

2) Sequential write of 800 packets of 512 bytes in a newly created file, which also yields a 400 KB file.

The size of 400 KB is chosen in order to allow the experiment to perform a sufficient number of calls to *jffs2_write_begin( )* and *jffs2_write_end( )*.

Results: The results are presented in Figure 6.12. The execution time of *jffs2_write_begin( )* varies between 5 and 30 µs. Therefore, it is a very fast operation, especially in comparison to *jffs2_write_end( )* discussed in the paragraph below. As a result, we do not linger on the *jffs2_write_begin( )* performance.

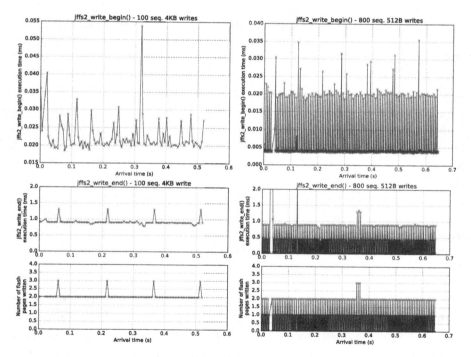

Figure 6.12. *Performance exploration of* jffs2_write_begin( ) *and* jffs2_write_end( )*: on the left, the results for 100 writes of 4 KB, on the right, 800 writes of 512 bytes. On top, the performance of* jffs2_write_begin( ) *can be seen, in the middle, the performance of* jffs2_write_end( )*, and at the bottom, the number of flash pages written for each* jffs2_write_end( ) *call*

Regarding *jffs2_write_end( )*, the execution time of this function strongly depends on the number of written flash pages. For requests of 4 KB, it can be seen that most times two flash pages are written. This corresponds to an execution time of a little less than a millisecond. At regular intervals, three flash pages are written. This is due to the phenomenon already encountered in the previous section dealing with read operations (see Figures 6.5 and 6.6). When three pages are written, an execution time of approximately 1.3 ms is observed for *jffs2_write_end( )*.

Regarding the write operations of 512 bytes, it is possible to reach the same conclusion in regards of the impact of the written flash pages. It can also be seen that a large number of calls to *jffs2_write_end( )* do not trigger any flash page write. This is due to the effect of JFFS2 write buffer; when a write within a Linux page is performed with a size smaller than the size of a flash page, if the amount of free space in the write buffer of JFFS2 is sufficient, data are stacked in this buffer. In this case, the execution time of *jffs2_write_end( )* is observed to be very short. We know that when the write buffer is full, a flash page is written. Therefore, a question might arise about why, in the experiment with requests of 512 bytes, several write operations of 2 or 3 flash pages are observed. This is due to the fact that when JFFS2 notices that the last byte of a Linux page has been written, whichever the size of the request, the entire Linux page is rewritten. Therefore, this yields a write operation of 4 KB of data in flash memory, which can translate into two or three write operations of flash pages (one flash page is equal to 2 KB on our hardware platform), depending on several factors such as compression and size of free space in the write buffer. This process takes place in order to reduce the fragmentation of a JFFS2 file system.

No difference in performance is observed when the write operations are performed at random positions in files. This is quite normal, considering that JFFS2 performs sequential write operations in flash memory, regardless of the write access mode at logical level. Random accesses to a JFFS2 file can be classified in two categories:

1) Random accesses *without overwriting*: these are random write operations on a newly created file with a zero size, or random accesses to an offset greater than the file's size. This is a rather peculiar type of access, which is why this work does not linger on it. In the Linux *man* page of the system call *lseek( )*, Linux defines the following management for this kind of

accesses: they are authorized and they trigger the creation of a "hole" between the position of the last byte of the file before the write operation and the position of the first written byte. A read of data in the hole yields a sequence of "0". This is illustrated in Figure 6.13. JFFS2 manages this phenomenon by writing a node for the file, with a small amount of metadata, representing the hole;

2) Random accesses which overwrite data in a file: these are update operations. There is no difference between write functions in terms of performance during such an access, however:

– these operations cause fragmentation, which depends on how logical data contained in the new nodes happen to overlap with the data of old, partially or totally overwritten nodes. Fragmentation affects read performance, as described in the previous section;

– invalid data are created, which lead to the execution of the garbage collector, whose impact will be described in the following paragraphs.

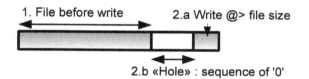

Figure 6.13. *Creation of a "hole" following a write access to an offset (or position) greater than the size of a file*

Regarding power consumption analysis, the following experiment is performed:

Goal: Investigating power consumption of write accesses in JFFS2.

Method: Sequential write operations of 1280 requests of 4 KB and 10240 requests of 512 bytes are performed on a file. As in the read experiment, the amounts of requests are great in order to obtain a readable power consumption curve. Therefore, the total size of the file after each experiment is equal to $1280 * 4096 = 10240 * 512 = 5$ MB.

Results: Figure 6.14 presents the evolution of the power consumption as a function of time, measured for the CPU and the memory chip of Omap board. Like in the read experiments, clearly visible levels of power consumption are observed during the accesses, and an average value of power, not influenced by the size of request, can be calculated.

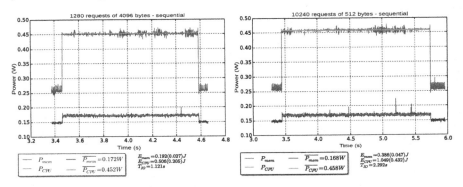

Figure 6.14. *Power consumption during the sequential write of files with 1280 requests of 4 KB each (left) and 10240 requests of 512 bytes each (right). For a color version of this figure, see www.iste.co.uk/boukhobza/flash.zip*

6.2.2.2. Garbage collector and initial state

The results presented here are issued from [OLI 12]. The garbage collector of JFFS2 is launched with the purpose of recycling invalid flash space by erasing blocks in order to create free space. Two operating modes can be distinguished for this process:

1) *Background operation of the garbage collector*: by means of a Linux kernel thread executed with a low priority, the background garbage collector recycles blocks during the I/O timeouts. As a consequence, this operating mode does not affect the response time of I/O operations.

2) Garbage collector execution below *critical free space threshold*: when a JFFS2 tries to write a node in flash memory, the threshold of free space is checked. If the free space is not sufficient, the garbage collector delays the write request which triggered the write of the node, because it has to be executed foremost. Therefore, this garbage collector affects the response time.

Although the background garbage collector, also called asynchronous garbage collector, does not produce an impact on performance, it increases power consumption. In order to observe this effect, the following experiment is performed:

Goal: Observing the impact on power consumption of the background garbage collector in JFFS2.

Method: A large number of invalid data are created in a JFFS2 partition. A file is created and written sequentially. Once the file is created, its data are overwritten by means of a sequence of write requests at random positions. The size of the file is equal to 5 MB, and the number of random write requests is of 1280, each with a size of 4 KB. Power consumption is measured during and after the experiment.

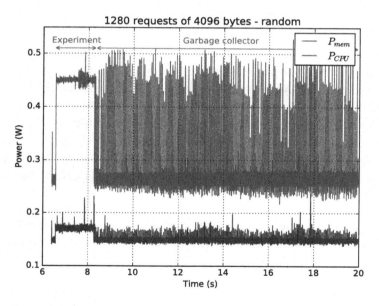

Figure 6.15. *Execution of the background garbage collector after an experiment generating a great amount of invalid data*

Results: The results are presented in Figure 6.15. After the experiment, a sequence of power consumption peaks corresponding to the garbage collector execution is observed, both in the CPU and the memory chip. Note that this operation lasts 40 seconds after the experiment (the end of this interval is not shown in the figure). These peaks correspond to the execution of garbage collection algorithms, and in particular to flash I/O operations performed for this purpose: read of still valid data in victim blocks, write of this data into the current block and erase of victim blocks. This behavior is confirmed by a Flashmon trace of the flash accesses during this phase of background garbage collection. Although the background garbage collector does not alter the performance of I/O user's tasks in a negative way, in the case of a fragmented file in flash memory the still valid nodes moved by the garbage collector can be merged into a new node (with a maximum size of 4 KB). The garbage collector plays thus the role of a defragmenter. The following rules are applied:

– the merged nodes have to be adjacent from the logical point of view;

– the merged nodes have to be situated within the same flash block, which is the victim block chosen by the garbage collector. Otherwise, as merging causes the merged nodes to be invalidated, the garbage collector would create invalid space, contrarily to its original purpose.

In order to verify this, it is possible to observe the time taken by a file read considered in the previous experiment. This file is read sequentially using *read( )* function. The read is performed (1) right after the random write, in other words before the garbage collector prompts a defragmentation, and (2) after the defragmentation. The time measured for (1) is equal to 0.83 seconds, while for (2) 0.70 seconds are measured, i.e. an 18% difference. Using Flashmon, it is confirmed that the number of read operations on flash pages is smaller in case number (2).

In the case of garbage collector executed *below critical free space threshold*, as mentioned previously, its execution directly affects the performance of the user's I/O tasks.

Goal: Showing that when the free space is very small, garbage collection causes an increase of write latency and a less efficient wear leveling.

Method: In order to observe this effect, at first, the following experiment is performed: within a JFFS2 partition of 100 MB, a file of 20 MB (20% of flash space of the partition) is created sequentially. This file is called file A. Then, a great number of write operations using the *write( )* function is performed in a loop on this file, by employing variable sizes at random positions. This creates invalid space and triggers the execution of the background garbage collector. Furthermore, in the same partition, a file B is created before the experiment, with a variable size: 30, 60 and 75 MB. This file contains valid and non-fragmented data. For this reason, during random accesses to file A, free available space is variable (according to the size of B). In particular, it is very small in the case when B has a size of 75 MB: $100 - 20 - 75 = 5$ MB. Therefore, in this case, an execution of garbage collector below critical free space threshold is expected. The experiment's process is illustrated in Figure 6.16. During random accesses to file A, the execution time of each *write( )* is measured. Additionally, the average number of erase operations per block of the partition, as well as the erase distribution, is measured using Flashmon.

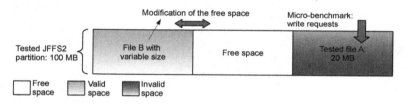

Figure 6.16. *Methodology of the experiment concerning the garbage collector execution below critical free space threshold*

Results: The results are presented in the curves on top and at bottom left of Figure 6.17. It can be seen that while the average execution time of each *write( )* call does not vary when a lot of free space is available, a higher average response time is observed when the fill rate is high (75 MB). This performance drop is due to the execution of garbage collector below critical free space threshold, which delays the write requests. Therefore, the number of erase operations is higher, and wear leveling is worse (see Figure 6.17).

In order to investigate the interactions between the two operating modes of garbage collector, an additional experiment is performed:

Goal: Exploration of interactions between the two operating modes of JFFS2 garbage collector.

Method: A JFFS2 partition of 20 MB is completely filled with a file, which is erased just before the start of the experiment, thus saturating the partition with invalid space. Then, a write of a new file is launched on this partition, by varying the request size (between 2 and 64 KB) and the time between each *write( )* request: 0, 10 and 100 ms. The average response time of each *write( )* call is measured.

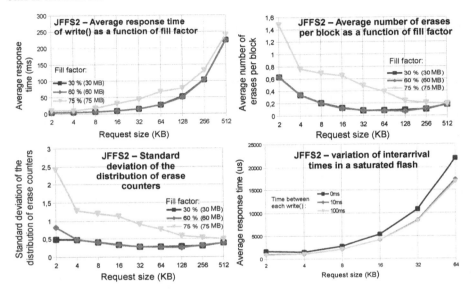

Figure 6.17. *Results of the experiment concerning the garbage collector execution below critical free space threshold (top and bottom left), as well as the results of the experiment of varying the arrival time during the execution of the garbage collector (bottom right)*

Results: The results are presented at bottom right of Figure 6.17. It can be seen that when the interval of time between each *write( )* is large (10 and 100 ms), the background garbage collector is able to perform its work and it brings the free space level above the pre-defined threshold, which avoids the execution of garbage collector below critical threshold. In fact, a higher average latency can be seen in the case of accesses without delays between write requests.

6.2.3. *Conclusion*

FFS file read/write operations and garbage collection performance and power consumption are mainly affected by the number of flash access performed during each of these operations. As our study focused on JFFS2, additional tests, not described here, confirm this observation for other FFSs supported by Linux (YAFFS2 and UBIFS). The parameters that affect the number of performed flash I/O accesses are the following:

1) For read operations, the accesses are subdivided by VFS into accesses to one or more Linux pages, which are read in loop using *readpage* function of the file system. An eventual fragmentation of the file nodes in flash plays an important role. On the contrary, a sequentially distributed file in flash memory benefits of the MTD read buffer, which reduces the number of flash accesses. The distribution of the file in flash memory is determined by the way the file was written (or updated), and by the defragmentation performed in the background by the garbage collector thread.

2) For write operations, the JFFS2 write buffer is employed for small-sized write tasks. Therefore, the number of written flash pages is determined by the amount of data written by the application together with the effect of the write buffer. Below a critical free space threshold, the garbage collector introduces additional latency for write operations, mainly due to the flash accesses corresponding to the recycling of still valid nodes.

6.3. VFS level

VFS level can be considered to be in direct contact with the application that accesses storage, as it is the layer that receives the system calls issued by the application. After being processed internally by VFS, its I/O commands are transferred to the FFS level, as it was described in the previous section. Like the driver and the FFS, VFS adds an overhead in performance and power consumption to the I/O operations. Page cache, which buffers the I/O operations on files, is maintained at the VFS level. Two mechanisms are associated to the page cache: a prefetch read mechanism, called *read-ahead*, and the *write-back* function, which buffers the write operations in the page cache and performs them at a later moment within the file system.

6.3.1. *Page cache*

All data written or read from files are placed into the page cache, and they remain there after the access, until their deletion. The deletion happens when the amount of free RAM becomes low and the processes being executed by the operating system demand more memory space. It is also possible to empty the page cache using the following command:

```
echo 1 > /proc/sys/vm/drop_caches
```

Every read access on data within the page cache is thus performed through the RAM. We seek to observe the impact of the page cache through the following experiment:

Goal: Observing the difference in performance between a read of a file from flash memory and a read of the same file from RAM memory (page cache).

Method: A non-fragmented file of 12 MB is created on a JFFS2 partition. The size of this file is deliberately large, because we want the experiment to act for a long time. In fact, a read from the page cache is a transfer from RAM, therefore, a rather quick one, and it is desirable to obtain a readable power consumption curve. The page cache is emptied and the file is read sequentially at first. This causes the data of the file to be loaded into the page cache. A second sequential read is then performed. Power consumption and execution time of these two operations are measured.

Results: The results are presented in Figure 6.18. Clearly, a read from the page cache is much faster than a read from flash memory, which is normal, because of the throughputs of these two types of memory. Reading 3000 Linux pages from the page cache takes 0.06 seconds, compared with 1.2 seconds from flash memory. Furthermore, during the accesses, a much greater power is observed in the CPU chip and the memory chip when a read is performed from the page cache, with respect to a read in flash memory. During the latter, an average power of 0.178 W is observed in the memory chip, and a value of 0.489 W on the CPU. As for a read from the page cache, an average power of 0.362 W is obtained for the memory chip and 0.569 W for the CPU. Nevertheless, as the execution time is much shorter during a read in RAM, the energy consumed by this experiment is significantly lower than in the case of a read in flash memory: 0.013 (memory) + 0.018 (CPU) = 0.031 joules against 0.037 + 0.268 = 0.305 joules (difference of one order of magnitude). Note that these energy values are calculated using the power values from which idle power values were subtracted.

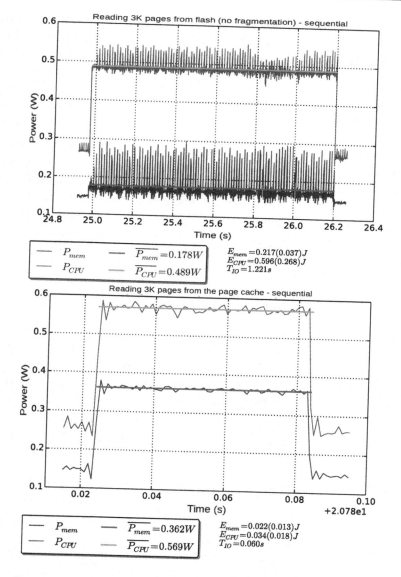

Figure 6.18. *Sequential read of 3000 Linux pages in a file from the flash memory (top) and from the RAM page cache (bottom). For a color version of this figure, see www.iste.co.uk/boukhobza/flash.zip*

Therefore, it is possible to conclude that, although power consumption is higher during RAM accesses, thanks to the speed of read operations in RAM it is more convenient, in terms of energy, to read from RAM rather than from flash memory. The page cache thus contributes to save energy. This is due in particular to the power consumption of CPU, which remains high during the transfers.

6.3.2. Read-ahead mechanism

The results illustrated here are obtained from [OLI 14b]. The *read-ahead* mechanism presents a significant impact on the performance of read accesses to files managed by FFS. The main interest of *read-ahead* mechanism, when used with a hard disk drive, is that these systems support asynchronous I/O tasks: while some data are prefetched from the file system, the CPU can execute other tasks. The first remark in the context of FFSs is that asynchronous I/O tasks are not supported by the NAND driver MTD. In order to verify that, a simple experiment is performed:

Goal: Verifying the synchronous execution mode of MTD.

Method: A file is read sequentially using a loop of *read( )* calls. The experiment is launched 3 times by varying the time between each *read( )* call, equal to 0.5 ms and 50 ms (via *usleep( )*). The execution time of each *read( )* call is measured using *gettimeofday( )* function. Then, the average execution time of a *read( )* call is calculated. The experiment is performed both with JFFS2 and YAFFS2 FFS[2]. The size of the file is fixed to 5 MB, and the number of *read( )* calls within the loop varies, following the power of 2, from 8 to 1024 calls. The size of a read (a *read( )* call) is equal to 4 KB.

Results: The results are presented in Figure 6.19. By tracing *read-ahead* calls in the kernel with VFSmon, it is observed that this mechanism is quite busy during the experiment. Therefore, an improvement of performance might be expected when the time between *read( )* calls is significant, which would make possible an asynchronous prefetch. However, it can be seen that the time between each call does not affect performance at all. This is due to the fact that the calls to MTD driver, and, as a consequence, the calls to *jffs2_readpage( )* and *yaffs2_readpage( )*, are synchronous. In other words, if

---

2 YAFFS2 has been added for the sake of completeness of the study presented in the article mentioned above.

a JFFS2 *read( )* call (with a size of a Linux page) triggers a prefetch of 31 other Linux pages from the FFS, then a read of 32 pages will be performed during the execution of the system call *read( )*, and it will block the calling process until all the 32 pages are present in the page cache. Because of this behavior, *read-ahead cannot affect the performance of FFSs in a positive way*. In the best case scenario, when all prefetched data are actually requested by the process, read-ahead does not affect performance at all. In other cases, when prefetched data are not used by the process (for example, a sequential read flow that stops before the end of a file), read-ahead affects performance negatively. In such cases, the energy expense of read-ahead mechanism is further increased due to the number of read Linux pages in excess.

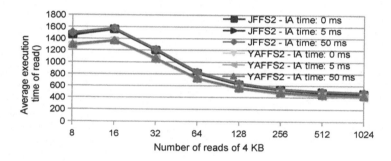

Figure 6.19. *Evolution of the execution time of* read( ) *calls as a function of their inter-arrival times. For a color version of this figure, see www.iste.co.uk/boukhobza/flash.zip*

We have modified the kernel code in order to deactivate read-ahead for JFFS2 and YAFFS2, and then we have tried to quantify the improvement caused by this deactivation.

Goal: Observing the improvement of performance due to the deactivation of read-ahead.

Method: The experiment consists in reading two files of 5 MB and 10 MB, in a sequential and a random way, in a loop with a number of *read( )* calls varying between 8 and 2048.

Results: The results are presented in Figure 6.20. It can be seen that the improvement of performance due to the deactivation of read-ahead is particularly significant in the sequential case and when the number of read pages is low. By tracing the flash I/O accesses with Flashmon, it is possible to confirm that the number of read operations on flash pages is actually lower when read-ahead is deactivated. An improvement of performance is also seen in the random case, which means that read-ahead is operating during this kind of access. Furthermore, it can be seen that, the greater is the amount of read data, the lower is the improvement of performance; this is due to the fact the prefetched data are actually used when the amount of data requested by the process is close to the size of the read file.

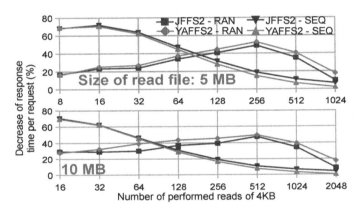

Figure 6.20. *Improvement of performance due to the deactivation of read-ahead*

6.3.3. *The write-back mechanism*

The write-back mechanism in the page cache is not supported by JFFS2 (neither by YAFFS2). In fact, the write operation is performed in flash memory during the execution of *jffs2_write_end( )* function. UBIFS, however, supports this feature.

Goal: Observing the impact of the write-back in the page cache on performance.

Method: By means of *write( )* function, a file is written sequentially in a flash partition (JFFS2, YAFFS2 and UBIFS), by varying either the number of write requests or their size. When the number of requests is varied, the size is fixed at 4 KB. When the size varies, that number is fixed arbitrarily at 500. The total execution time of the write loop is measured.

Results: The results are presented in Figure 6.21. It is observed that for JFFS2 and YAFFS2 the execution time curves follow a similar evolution behavior with respect to the amount of written data. As for UBIFS, it presents considerably shorter execution times, thanks to the support of the write-back mechanism in the page cache. Actual write operations in flash memory are not performed during the execution of *write( )* calls. Data are updated only within the page cache (in RAM), which helps accelerating the operation in a significant way.

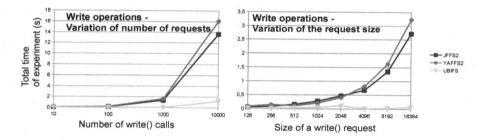

Figure 6.21. *Impact of the write-back mechanism of the page cache, operating only in UBIFS. When the number of requests is varied, the size is fixed at 4 KB. When the size varies, the number is fixed at 500. Logarithmic scale for x-axis*

6.3.4. *Conclusion*

The page cache and the associated mechanisms strongly affect performance. Concerning power consumption, as for the FFS level, it is possible to identify values of average power, in particular regarding RAM accesses due to page cache hits in read operations. Although the power during RAM accesses is higher than what observed during flash accesses, the total energy consumed by the operations depends mainly on their execution time.

6.4. Conclusion

In this chapter, we presented the different elements affecting performance and power consumption of storage systems based on embedded flash memory, managed by FFS on Linux. We focused in particular on the context of read and write operations on data within files.

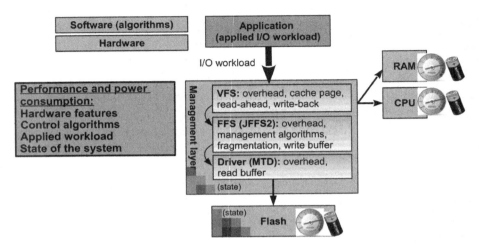

Figure 6.22. *Diagram of different elements affecting performance and power consumption in a storage system based on flash memory*

The various operational mechanisms that process the I/O application workload at all levels of the flash storage management stack are: the *virtual file system*, the *file system* itself (FFS) and the *NAND driver*. Each of them defines the I/O accesses performed on the flash chip and the behavior of the different RAM cache buffers implemented at all levels. The actual impact of these elements on performance and power consumption is shown using a set of experiments. These experiments have been performed on a hardware platform, the *Mistral Omap3evm* board. The experiments employ previously described tracing tools and the *Open-PEOPLE* remote platform for power consumption measurements. In summary, a generalization can be done by stating that power consumption and performance of a memory-based storage system is the result of interactions between several elements (see Figure 6.22) which are:

– *Characteristics of hardware elements* composing the storage system: for example, latency of basic flash operations or the idle or active power on CPU, flash and RAM chips.

– *Algorithms of management layers* that process the workload. Each layer (VFS, FFS, MTD) adds an overhead in terms of time and energy. For each layer, the following remarks can be done:

- at the driver level, MTD makes possible the execution of basic operations on flash memory. A read buffer is present at this level, and it can affect performance and power consumption in the case of a temporal locality;

- at the FFS level, we have focused on JFFS2. It was observed that the way in which the file is distributed in flash memory (nodes fragmentation) can strongly affect performance and power consumption of read operations. Regarding write operations, the presence of a buffer absorbs write operations with a size smaller than a flash page size;

- at the VFS level, page cache buffers all the I/O accesses and can provide repeated read operations of the same Linux pages from the RAM, which strongly affects performance and power consumption depending on the I/O workload of the application. Associated with the page cache, read-ahead and the write-back mechanisms perform a prefetch (synchronous in the case of FFSs) and delayed write operations from the page cache (only in UBIFS), respectively.

– *Characteristics of the I/O workload of the application.*

– *State of the system*: state of caches, of flash memory (amount of free, valid or invalid space), of fragmentation of a JFFS2 file, etc.

PART 3

Flash Translation Layers

Experience is a memory, but the opposite is true
Albert Camus

With memory one overcomes all
Alfred De Musset

7

Flash Translation Layer

This chapter provides an overview of the achieved state of the art of FTLs (or *Flash Translation Layer*). As a reminder, an FTL is a hardware/software layer located in the controller of the storage device that makes use of flash memory such as SSDs, USB keys or SD cards. After a short introduction, we will address the following points:

1) Basic mapping schemes from which the totality of FTLs are built.

2) More complex state-of-the-art mapping schemes making it possible to address important concepts related to the performance of these systems.

3) Wear-leveling mechanisms that allow balancing the wear out on flash memory blocks.

4) Garbage collection mechanisms providing a means for recycling invalid space;

5) Cache systems specific to flash memory designed to maximize the lifetime as well as the performance of flash devices.

7.1. Introduction

As we have described in Chapter 2, integrating flash memory in a computer system involves the management of a number of constraints: erase-before-write, the difference in granularity between reads/writes and erase operations as well as the limited number of write/erase cycles that a given memory cell can sustain. These constraints require the implementation of specific management mechanisms. These can be implemented either entirely in software on the host system via dedicated file systems such as those mentioned in Chapter 3 or in the hardware, that is to say integrated into the controller of the flash memory on the device being considered. In the latter case, this is referred to as an FTL (for *Flash Translation Layer*).

The main features of FTLs that will be described in this chapter are the following:

– Physical to logical mapping: due to the erase-before-write constraint, the modification of data implies that the latter be rewritten into another page. Therefore, it becomes necessary to keep track of the location of the data. This is achieved through mapping schemes;

– Garbage collection: updating the data involves mainly two operations: (1) copying the data into another page and (2) the invalidation of the initial copy of the data. Over time, invalid pages tend to accumulate due to the numerous modifications that may occur. In order to recover the free space for future writes, a garbage collector scans the flash memory and recycles entire blocks;

– Wear-leveling algorithms: as indicated in Chapter 2, a flash memory cell can undergo a limited number of erase/write cycles due to temporal and spatial localities observed in several I/O application workloads, it is necessary to make available a mechanism that provides a means to level the wear out in a uniform manner in all the blocks of the flash memory in order to maintain the function of the whole flash space for as long as possible.

7.2. Basic mapping schemes

The process of mapping consists of the translation of the address originating from the host (the application or more accurately the device driver at the system level), which is called a logical address, into a physical address. This translation involves a mapping table that is managed by the FTL layer. Basic mapping schemes are related to the granularity with which the mapping is performed. These schemes can be classified into three large families: page-level mapping, block-level mapping and finally a family of hybrid mapping schemes (page and block-level).

7.2.1. *Page-level mapping scheme*

In the page-based mapping scheme [BAN 95], each logical page address is mapped to a physical page address, and this is regardless of the other pages of the same block. This mapping scheme is very flexible and results in very good performance. The major constraint of this type of scheme is the size of the mapping table, which depends on the number of pages of the flash memory and can be very significant. As a result, such a mapping table can hardly be maintained in the RAM memory of the FTL (SSD controller). For example, if there are a number of blocks equal to B and each block is composed of P pages, the total number of entries in the mapping table is equal to $P * B$. We consider a numerical example to establish the orders of magnitude; if a flash memory of 32 GB is made available, with 2 KB per page, and if each entry in the page mapping table is 8 bytes, this table will have a total size of 128 MB, which is too large to be stored on on-board RAM.

Figure 7.1 shows an example of this type of mapping scheme.

7.2.2. *Block-level mapping scheme*

The main idea upon which block-based mapping relies [SHI 99] is to consider the block granularity rather than that of a page. Therefore, the logical address of a page consists of a block number and of an offset within this block. The offset is not altered by the process of mapping (see Figure 7.1). If considering the same example as that of page-level mapping, the number of entries will be equal to the number of blocks, which is B and therefore, instead of 128 MB, the table will have a maximum size of 2 MB. In addition, we need less space per entry because block numbers can be encoded by making use of fewer bits.

Even if block-based mapping is more realistic, when considering the size of the mapping table, it has nevertheless a real downside. In fact, when a page has to be updated, the write must be carried out within the same offset in the block, as a consequence, the whole block containing the page must be copied into a new free block. This amounts to copying the set of all valid pages in addition to the newly modified page, and finally to invalidate the block previously used. This operation is extremely costly and thus unrealistic.

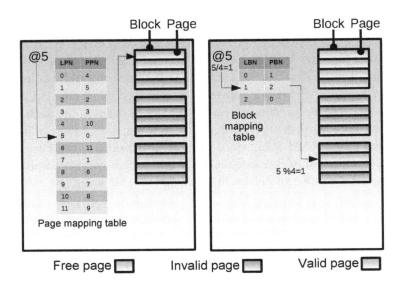

Figure 7.1. *Page-level and block-level mapping schemes. The figure on the left shows an example of a page mapping scheme when accessing page number 5. The figure on the right shows an example of a page-based mapping scheme when accessing the same page number. In this case, two operations are performed: an integer division operation allowing the block number to be obtained and a modulo operation to get the page number in the block. LPN represents the logical page number, PPN, the physical page number, LBN, the logical block number and PBN, the physical block number. For a color version of this figure, see www.iste.co.uk/boukhobza/flash.zip*

7.2.3. Hybrid mapping scheme

Several state-of-the-art hybrid mapping schemes have been proposed in order to address the problems of the two previous schemes, namely the size of the mapping table and data update operation cost. These hybrid schemes make use of both types of mapping previously introduced. A large number utilize a global block-based mapping and maintain a page mapping table for a subset of pages for which block mapping would be expensive. As a result, the block-based mapping table and the page-based mapping table which have a small size are kept in RAM. An example of this type of mapping is shown in Figure 7.2.

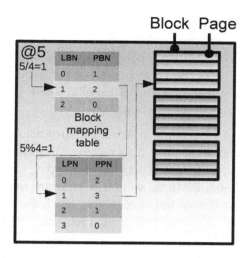

Figure 7.2. *Hybrid mapping scheme in which the main mapping table is a block-wise table. In addition, for some blocks (for example here LBN 1 / PBN 2), a page-based table mapping is used. For a color version of this figure, see www.iste.co.uk/boukhobza/flash.zip*

7.3. Complex mapping schemes

As introduced in previous sections, a good mapping scheme should have a small memory footprint while maintaining good writing performance.

There are three main types of schemes which are derived from the basic schemes and are capable of addressing the problems of the size of mapping tables and of the performance of writes:

1) If we reconsider the page-level mapping scheme, which shows good performance with writes because there is no need to copy other pages of a block during an update, the problem of the size of the mapping table can now be solved by simply keeping part of the latter in RAM. The rest of the table will be stored in flash memory and loaded into RAM when needed. Several FTLs are based on this principle including DFTL [GUP 09] and SFTL [JIA 11];

2) If we consider the block-level mapping scheme, in order to minimize the cost of data updates, a possible solution is to allocate pages or an additional block to a given data block. These additional pages or these blocks are called *log* pages or *log* blocks. These log pages and blocks are employed to absorb

the updates of pages without having to update the entire block (we will give more details about this scheme further in the text). The data in log pages or log blocks are merged with data from data blocks later in a single pass, which prevents additional copies. Some examples of this type of scheme will be provided later. As an example, FAST [LEE 07], BAST [KIM 02] and LAST [LEE 08] are FTLs that fall into this category;

3) Another way of proceeding consists of partitioning the flash memory in two asymmetrical areas, a small area managed in page-wise fashion and a more significant area block-wise. This type of mapping scheme aims to write the pages very often modified in the section in which the mapping is performed page-wise, while data that do not change very much are stored in the block-wise mapping section. Two examples of such schemes are: WAFTL [WEI 11] and MaCACH [BOU 15].

7.3.1. *Log block FTL*

In this section, we will describe FTLs representative of state-of-the-art.

7.3.1.1. *Mitsubishi*

Mitsubishi's FTL is a block-based mapping FTL [SHI 99]. This is one of the first to use the concept of log space in order to absorb page updates. Indeed, in this FTL, each data block contains a few additional pages employed as log space. These are pages not visible to the application space and are used as soon as a data update is performed in one of the pages of the block. It is only once all of the log pages are written that the following update operation generates a copy operation to a new free block (see Figure 7.3). This FTL makes it possible to reduce block copies at the cost of additional flash storage inserted in each block. As an example, if 8 log pages are added to each block containing 64 data pages, the addition in terms of space represents 12.5%.

7.3.1.2. *M-Systems*

In the M-Systems FTL [BAN 99] rather than having a few log pages inside each block, the idea is to have entire log blocks. This means that for a logical block of data, there are two physical blocks, a data block and a log block that will absorb the updates. Two different schemes are possible for this FTL: ANAND and FMAX. In the first, ANAND, simpler in its implementation, when an update occurs with a page of the data block, the latter is achieved in the page with the same offset in the log block. It simply means that when it is desirable to modify the same page twice, the system is forced to perform a merge operation. This operation consists of merging valid data from the data block with those of the log block into the same free block (see Figure 7.3, example 2).

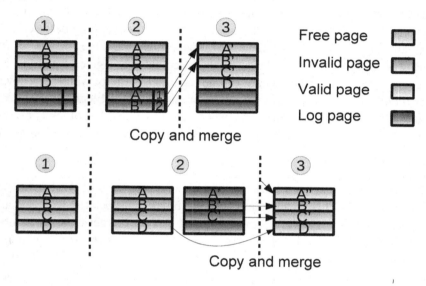

Copy and merge

Copy and merge

Figure 7.3. *Diagram showing two examples of using log pages in the first and log blocks in the second. In the first example, the following sequence of data is written: A, B, C, D, A', B', C'. For the first 6 writes (steps 1 and 2), due to log pages, even after the updates of A and B (A', B'), no erase operation takes place. On the other hand, as soon as C is modified (step 3), valid pages have to be copied into a new block. For the second example, the data sequence written is: A, B, C, D, A', B', C', A". After the first 4 writes the block is full (step 1) and at the first additional modification a log block is allocated. The latter can absorb the updates of A, B and C (A', B', C' in step 2). Finally, when A is updated for a second time (step 3), a merging operation between the data block and the log block is carried out*

The advantage of this version of the FTL is that there is no additional mapping to maintain between data blocks and log blocks. In FMAX, when an update operates on a page, this latter is copied to the log block at the first available location and the page number (of the original data block) is copied into the out-of-band (OOB) area; thus, it becomes possible to update the same page several times. The merge operation is done when there is no more space in the log block. The cost of this technique is an additional mapping table in order to know on which page of the log block a data block page has been last updated.

7.3.1.3. BAST or Block Associative Sector Translation FTL

BAST [KIM 02] is an FTL similar to FMAX but that limits the number of blocks used as log. This allows the management of the mapping between data block and log block pages by means of a page mapping table (instead of the OOB region). In BAST, a single log block is dedicated to a given data block. A merge operation is activated when: (1) the log block is full or when (2) a newly written block page is updated while there are no more log blocks available. In the latter case, a merge operation between data block and log block is carried out in order to recycle a block that will be able to be used again as a log block.

7.3.1.4. FAST or Fully Associative Sector Translation FTL

The main problem with BAST is the associativity level of the log blocks, which is set to 1. This means that every log block is associated with one and only one data block. However, if a significant number (larger than the number of log blocks) of data blocks undergo updating, the system will generate several merges with log blocks rarely used. The purpose of FAST is to avoid this problem by allowing a log block to be associated with several data blocks. In FAST [LEE 07], log blocks are divided into two regions, each region containing a block dedicated to sequential updates. In effect, if a block is completely updated, it is very disadvantageous to distribute the pages over different log blocks because the cost of merging would be very significant. Therefore, pages sequentially written are placed inside the same block. The second region contains the rest of log blocks that will absorb random updates in a completely associative way. This makes it possible to maximize the use of log blocks by minimizing the number of merging transactions. In FAST, log blocks are managed by a page mapping table.

7.3.1.5. LAST or Locality Aware Sector Translation FTL

FAST effectively manages to reduce the number of merge operations by maximizing the usage rate of log blocks. On the other hand, the cost of each merge operation becomes more significant since the log pages from a single block of data are scattered over several log blocks. When the system has to recycle a log block, one data block merging operation is thus no longer sufficient. As a matter of fact, it is necessary that the same number of merge operations be performed as there are different block pages in the log block. For instance, if a log block contains pages originating from two data blocks, 2 merge operations would have to be performed (one for each of the blocks) in order to recycle a single log block (see Figure 7.4). This is more accentuated if one takes into account the different access patterns that can be achieved with the data of different log blocks. Nonetheless, if we mixed very infrequently modified data (cold data) and very frequently modified data (hot data), unnecessary data movement would be created. Indeed, cold data would move from a log block to another without having been modified. The purpose of LAST [LEE 08] is precisely to reduce these data movements and thus to minimize the cost of merge operations. To this end, LAST first increases the number of log blocks for sequential access. It relies on the size of the requests to send updates to sequential log blocks. Second, LAST partitions the region of random log blocks into two, one hosting data very frequently modified (so-called hot data) and the other, the less frequently modified data (known as cold data). This allows the costs of merge operations to be minimized.

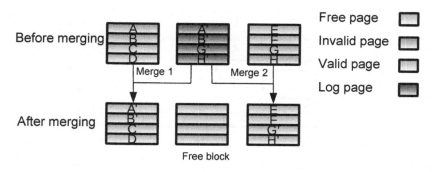

Figure 7.4. *Merging from two data blocks in FAST, two merge operations are necessary in order to release a single block. For a color version of this figure, see www.iste.co.uk/boukhobza/flash.zip*

Several FTLs endeavor to address the performance problems of FAST in addition to those already mentioned. As an example, KAST [CHO 09] for *K-Associative sector mapping FTL* is an attempt to reduce the cost of merge operations by limiting associativity and therefore the number of data blocks associated with a given log block. HFTL [LEE 09a], for *Hybrid FTL*, strives to address the problem of temporal locality in FAST by separating hot and cold data into different log blocks and by using a page-based mapping scheme for hot data and block-based for cold data. Several other more recent schemes do exist, such as BlogFTL [GUA 13] or MNFTL [QIN 11].

7.3.2. *Page-level mapping FTL*

As previously seen, the implementation of the conventional page-level mapping scheme is not an option because of the size of the table that has to be kept in RAM. Several studies attempt to address this problem, we will mention some of them in this section.

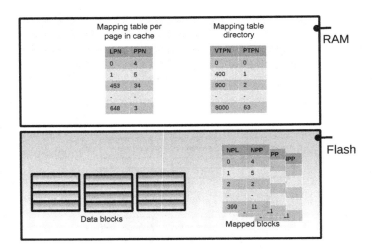

Figure 7.5. *Structure of the mapping tables in DFTL. The main mapping scheme is a page-based scheme of which part (the most recently accessed) is stored in RAM and the rest in flash memory. In addition, a directory of mapping tables in RAM is maintained in order to be able to find the flash block that contains a particular entry in the mapping table (MVPN and MPPN: the virtual translation page number and the physical translation page number, respectively). Flash memory is subdivided into two parts, a small section dedicated to the storage of the mapping table and the rest for the data*

7.3.2.1. *DFTL or Demand-based FTL*

DFTL is a purely page-based mapping FTL. In order to reduce the size of the memory storage of the mapping table, part of the latter is stored in flash memory and the most frequently accessed part of it is kept in RAM (see Figure 7.5).

CDFTL [QIN 10] proposes a two-level cache mechanism. It improves DFTL by taking into account the temporal and spatial locality of I/O workloads. Furthermore, only temporal locality is taken into account by DFTL by using an LRU algorithm for removing parts of the page table in flash memory.

7.3.2.2. SFTL or Spatial Locality Aware FTL

SFTL [JIA 11] is another page mapping FTL, which makes use of the sequentiality of I/O workload in order to reduce the size of the mapping table. If the I/O workload is sequential, it is not necessary to load several different references of the mapping table into RAM. Only one reference is enough, in addition to a value specifying the number of the following sequential references.

7.3.2.3. CFTL or Convertible FTL

CFTL [PAR 10a] focuses on correcting the defects of DFTL, namely poor performance of read operations due to the latency of the mapping scheme and to not accounting for spatial locality. CFTL is an adaptive FTL that stores data using a mapping scheme that depends on the access pattern. Data to which intensive read requests are applied employ block-wise mapping while very frequently updated data use page-level mapping. CFTL uses a cache scheme to store only part of the mapping tables (page and block-based), this scheme taking into account temporal and spatial locality. Although this type of FTL makes use of two types of mappings and is therefore hybrid, we have classified it among page-based mapping FTL because this latter scheme is thereof the principal scheme.

7.3.3. FTL with partitioned flash memory

The third family of mapping schemes is hybrid. The objective here is to partition the memory flash space into two regions, one small region involving page-based mapping and the other more significant region involving block-based mapping. We will provide two examples of these schemes.

7.3.3.1. WAFTL or Workload Aware FTL

WAFTL [WEI 11] is an adaptive FTL that stores data according to their I/O access pattern. In this case, flash memory is partitioned into two regions, one containing the data to be managed page-wise called *Page Mapped blocks* (PMB) and the other containing the data to be managed in a block-wise fashion called *Block Mapped Blocks* (BMB). WAFTL stores the data randomly accessed and partial block updates in the PMB area, whereas the BMB area stores the data accessed sequentially in addition to the mapping tables. WAFTL also makes use of part of the flash memory as a buffer so as to temporarily store sequentially written data before copying them into the BMB space. In addition, WAFTL maintains only part of the page-based mapping table in RAM.

7.3.3.2. *CACH-FTL or Cache Aware Configurable Hybrid FTL*

CACH-FTL [BOU 13a] is another example of FTL that considers flash memory as a partitioned area, one region using a block-level mapping scheme and the other region, of smaller size, utilizing a page-level mapping scheme (same principle as the BMB and PMB in WAFTL). CACH-FTL is based on the principle that, on the one hand, all SSDs are equipped with a cache in DRAM making it possible to speed up the access to the flash memory and, on the other hand, that most flash memory caches achieve flushing (eviction) involving groups of pages in order to minimize the merge operation between valid data in cache and valid data in flash memory for a given block (see further section 7.6). From this point, CACH-FTL is guided by the number of pages originating from the cache from each flush operation: if this number is below a certain configurable threshold, the pages are sent to the mapping area on a page basis. However, if this number is greater than this threshold, the pages are sent to the block-wise mapping area. In addition, in CACH-FTL a garbage collection mechanism allows data to be moved from the page area to the block area in order to recycle the space. Another more adaptive version of CACH-FTL has been proposed, named MaCACH [BOU 15]. The latter makes it possible to reduce the cost of garbage collection by dynamically modifying the value of the page flushing threshold according to a PID (*Proportional Integral Derivative*) feedback control system. This mechanism in turn enables that the limit value be adjusted depending on the filling rate of the page-based mapping region.

7.4. Wear leveling

As described above, each flash memory cell supports only a limited number of write/erase cycles. Beyond this limit, the cell becomes unusable.

The purpose of wear-leveling techniques is to try to keep the entire flash memory volume usable as long as possible. In order to achieve this, the FTL must balance the wear over the entire set of flash memory blocks.

In this chapter, we will use the same differentiation as [KWO 11] between wear leveling and garbage collection. Wear leveling is relative to block management, i.e. the choice of a block to use within free blocks and the garbage collector is responsible for the page management during the recycling, as we will see later. As discussed in Chapter 2, these two techniques are closely related and can be implemented in a single system.

Wear-leveling algorithms are based on the number of erase operations performed on a given block. If this number is greater than the number of average erase operations per block, the block is said to be hot otherwise it is said to be cold. Wear-leveling algorithms aim to keep the gap between hot and cold blocks as small as possible. This can be achieved by moving data from hot blocks towards cold blocks and vice-versa. Unfortunately, this operation is costly and therefore cannot be run frequently. As a matter of fact, there is a trade-off between wear leveling and performance.

We can group wear-leveling algorithms into two categories, those based on the number of erase operations and those based on the number of writes.

7.4.1. *Wear leveling based on the number of erase operations*

In these mechanisms, the number of erase operations is maintained for each block and is stored in the mapping table or in the out-of-band area of the flash memory. Obviously, it is preferable to have these metadata stored in RAM for performance reasons. The wear-leveling mechanism is launched as soon as the difference related to the number of erase operations becomes significant.

One of the techniques based on the number of erase operations, called *dual pool*, is described in [ASS 95]. This technique refers to two additional tables that are used in addition to the conventional block-based mapping table: a table of hot blocks and a cold block table. The hot blocks table contains the list of references of blocks that have been highly solicited and sorted by descending order in terms of number of erase operations undergone, while the cold blocks table contains the references of less erased blocks. With the *dual pool* algorithm, whenever the system requires a free block, a reference is taken from the list of cold blocks. Periodically, the system recalculates the average number of erase operations per block and rebalances both reference tables.

Some wear-leveling mechanisms operate only on free blocks while others also operate on blocks containing data [ASS 95, CHA 05]. For instance, if a block contains seldom modified data (cold or also called static data), these data are copied into blocks that have undergone several erase operations.

7.4.2. *Wear leveling based on the number of write cycles*

These algorithms maintain the number of writes that are applied to a block or to a set of blocks. In [ACH 99], in addition to the number of erase operations, a counter of the number of writes is kept in RAM and updated at each write cycle. The proposed algorithm consists of storing the most frequently modified data in the less erased blocks, and conversely the less frequently modified data are stored in the most frequently erased blocks. The technique discussed in [CHA 07] is very similar to that previously seen [ASS 95] but makes use of the number of writes per block. Two tables are maintained according to the number of writes carried out in the blocks. The algorithm interchanges data between regions that are frequently utilized and those that are less frequently utilized.

In order to decrease the size of the tables used for wear leveling (and therefore their memory footprint), some algorithms consider groups of blocks rather than individual blocks [KWO 11].

7.5. Garbage collection algorithms

The garbage collector mechanism (or *Garbage Collection*) allows the FTL to recycle invalid space into free space. According to [CHI 99a], a garbage collector must answer several questions:

– When should the garbage collector algorithm be launched?

– What blocks should the garbage collector be concerned with and how many should be used?

– How should valid data in victim blocks be rewritten?

In order to complete the operation, we must answer the last question related to the block that must be chosen for rewriting the valid data; it is part of the job of the wear-leveling mechanism. Garbage collection can be coupled to a wear-leveling mechanism within the same service as in the JFFS2 file system [WOO 01].

A garbage collection mechanism must minimize the cost of recycling while maximizing the recycled space, without having a significant impact on application performance. The cost of recycling includes the number of erase operations carried out in addition to displacing the valid pages.

Garbage collectors are usually started automatically when the number of free blocks falls below a predefined threshold. However, it can also be launched during I/O timeouts, which allows space to be recycled without impacting response times. One of the most important metrics taken into account during the choice of the block to be recycled (also called the victim block) is the rate of invalid pages in the block (or group of blocks). To this end, the garbage collector should consider the state of each page, which is very expensive in terms of resources. Some garbage collectors keep the page state in a RAM while others make use of the out-of-band (OOB) area in flash memory.

Intuitively, we could think that the victim block must be the one containing the greatest number of invalid pages. Actually, this is true if the flash memory is accessed in a uniform way; however, the number of invalid pages is always taken into account but this is not the only criterion. In [KAW 95], the percentage of invalid pages denoted $1 - u$ is taken into consideration, in addition to the time elapsed since the most recent modification (denoted age in the equation). The proposed garbage collector relies on the following calculation of the score: $age * (1 - u)/2 * u$. The block having the most significant score is chosen as victim. In fact, this score does not only consider the rate of invalid pages in the block but also the age. In other words, if the block has been very recently modified, the system will probably not choose it as victim block, but it will give time so as to potentially be updated more often. As a result, more pages will probably be invalidated, or even all will, and in this case the block will be able to be directly erased. The coefficient $2 * u$ relates to the cost of recycling: a u for reads and another for valid pages writes.

In another garbage collection policy, called *Cost Age Times* or CAT [CHI 99a], hot blocks (frequently modified blocks) are allocated more time in order to have more chances of increasing their number of invalid pages. The equation, upon which is based this garbage collection for the computation of the score, is the following: $Cleaning_{cost} * (1/age) * Number_{cleaning}$, where $Cleaning_{cost}$ is equal to $u/(1 - u)$. Here, u is the percentage of valid data. The variable $Number_{cleaning}$ is the number of erase operations generated. CAT recycles the blocks with the lowest score.

The temperature (or access frequency) is also taken into account by garbage collection algorithms. In fact, it is relevant to separate the hot data from cold data. If hot data are isolated in a block and all of these data are updated, the recycling operation with this block is limited to an erase operation. Whereas if this block contains hot data as well as cold data, during recycling, cold data will have to be copied into another block thereby increasing the cost of recycling.

Several contributions are based on this separation between hot and cold data [CHI 08, CHA 02, SYU 05, HIS 08]. For example, in DAC (*Dynamic dAta Clustering*) [CHI 99b], the flash memory is partitioned into several sections depending on the frequency of data updates.

7.6. Cache mechanisms for flash memory

In order to optimize the performance of write operations, several cache mechanisms[1], placed at the level of the flash memory controller, have been proposed. These caches essentially make two optimizations possible:

1) Certain write operations are absorbed at the cache level, which avoids their being copied onto flash memory. This helps reduce the impact of the I/O workload on the lifetime in addition to the derived performance gain;

2) Write operations are buffered so that they are more easily reordered before sending them to the flash memory. This allows the overall write cost to be decreased.

---

1 It should be noted that this is not related to a cache making use of processor SRAM technology but to buffer memory between the flash memory and the host. It is usually DRAM technology that is being utilized.

In this section, we give a few examples of state-of-the-art cache mechanisms for flash memory.

CFLRU (Clean First LRU) [PAR 06a]: this algorithm takes the asymmetry of read and write operations into account in the cache page replacement policy. CFLRU makes use of an LRU list divided into two regions: (1) the working region containing recently used pages, and (2) the clean first region containing candidate pages to be evicted. CFLRU flushes out the first page that does not generate any write operation into the flash memory in the clean first region. If this proves impossible, it chooses the first page according to the LRU algorithm. This algorithm adds a buffer to pages that were read and written and every time a page is accessed, it is relocated to the beginning of the LRU queue.

FAB (Flash Aware Buffer) [JO 06]: this algorithm takes into account page groups belonging to the same block, and not just like pages CFLRU. FAB utilizes an LRU algorithm. Every time a page of a group is accessed, the whole group is placed at the beginning of the list. During a flush, FAB chooses the group containing the most pages. In the event that several groups of the same size do exist, FAB makes a choice according to the LRU order. The choice of the group that contains the largest number of pages makes it possible to minimize the reading of valid pages from the flash memory and thus to reduce its cost.

BPLRU (Block Padding Least Recently Used) [KIM 08]: BPLRU is a cache that makes use of the following three techniques: (1) implementation of a block LRU list for groups of pages belonging to the same block, (2) a page-padding technique in which the valid pages are read from flash memory into the cache before a block is flushed out and (3) LRU compensation: this last operation consists of moving the blocks written sequentially directly to the end of the LRU list. It is assumed here that these blocks written sequentially are only very rarely reused.

C-Lash (Cache for flash) [BOU 11b]: the basic idea of this cache algorithm was to propose a cache mechanism that replaces wear leveling and garbage collection. This mechanism has been used later with FTL as CACH-FTL and MaCACH. The idea behind C-Lash is to use two separate buffers, the first enabling written pages to be stored, and the second which maintains an LRU list of groups of pages belonging to the same block. The page buffer has the ability to preserve recently accessed pages. Once this buffer is full, C-Lash chooses the largest set of pages belonging to the same block in this region and moves them to the block-wise region (see Figure 7.6). When the block region is full, C-Lash starts flushing to the flash memory following the LRU algorithm.

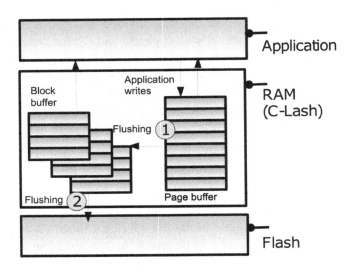

Figure 7.6. *C-Lash cache structure. This cache consists of two buffers, one that stores pages and the other groups of pages belonging to the same block. When the application performs a write, data are written into the page buffer, also called the p-space. If the latter is full, a flush of the larger set of pages belonging to the same block is performed in the block buffer (Flushing 1 in the figure), also called the b-space. If it is full, the choice of the block that is flushed out in the flash memory is made according to the LRU algorithm (Flushing 2 in the figure)*

There are other cache algorithms, such as PUD-LRU [HU 10], LB-Clock [DEB 09] and BPAC [WU 10]. It should be noted that most of these algorithms consider the granularity of page groups to optimize write operations into flash memory and utilize the concept of temporal locality based on LRU-type algorithms.

7.7. Conclusion

This chapter introduces the basic services of an FTL, namely: address mapping, wear leveling and garbage collection. We have also discussed cache mechanisms that are specific to the constraints of flash memory. In regards to mapping schemes, we have first started by describing basic schemes such as: (1) page mapping, (2) block mapping and (3) hybrid mapping. The first implies a translation table that is too large but with very good performance due to its flexibility (each page can be processed independently of each other). The second scheme has a major defect related to the performance of the updates: each data modification requires that the totality of the valid data be rewritten into the block in which it is contained at another location. And finally, the last scheme allows that the benefits of the first two be exploited. We have also briefly described several mapping schemes considered as state of the art: BAST, FAST, LAST, DFTL, WAFTL, CACH-FTL, etc. These hybrid translation or page-based schemes may present very different designs which can lead to differences in terms of performance (as we will see in Chapter 9).

We have subsequently described some wear-leveling algorithms that provide the means to manage the wear over the entire surface of the flash memory, and giving a few examples based on the number of erase operations and the number of writes. Next, we have described the main parameters taken into account by garbage collectors as well as some examples of the state of the art. These parameters are generally the proportion of invalid pages in a block, the date of the last update and the cost of the garbage collection operation. Finally, we have described the objectives of caches in devices relying on flash memory with some examples presented such as CFLRU, BPLRU and C-Lash.

The FTL-based storage device considered in this book is the SSD. In the following chapter, we describe a methodology to study the performance and energy consumption of storage devices and in particular of SSDs.

8

Methodology for the Evaluation of SSD Performance and Power Consumption

The objective of this chapter is to describe a methodology for measuring performance and power consumption that has been used to characterize the behavior of SSDs and their impact on a computer system. The methodology exposed in addition to the material employed in the writing of this book is obviously not the only way for the reader to understand the measurement of energy consumption of storage systems. Nevertheless, it provides an idea or even an interesting case study of the analysis of energy consumption in the context of data storage.

This chapter is composed of five sections:

1) An introduction to the methodology for measuring performance and energy consumption for which we have opted.

2) The description of the I/O software stack in Linux (supplements to the previous chapters).

3) The context in which the measurement methodology has been developed.

4) The description of tracing tools used for I/O characterization.

5) The description of the methodology for power consumption measurement.

8.1. Introduction

We have described in Chapters 5 and 6 a methodology targeting the analysis of performance and energy consumption as well as the results of the application of this methodology in an embedded system. The methodology was specific to the integration of flash memories with a software management layer (dedicated flash file system). In this chapter, we describe the methodology enabling the exploration of performance and energy consumption of flash memories with management layer integrated in the controller (FTL). We shall focus on SSDs. Indeed, they are largely integrated within current computing systems (for example: PC desktops, laptops, servers and data centers). We will thus analyze the impact of these storage systems on performance and power consumption of a computer system.

Several elements justify the use of a different methodology than that outlined in Chapter 5, among which:

– the fact that the FFS be Open Source and that FTLs are opaque and can be considered as black boxes;

– the software stack of the kernel managing I/Os is different in both, which is addressed in this chapter;

– hardware platforms are different and require dissimilar measurement tools.

8.1.1. *Method and tools*

In order to explore performance and energy consumption of flash storage systems with integrated management layer (SSDs in our case), we will mainly rely on performance measurement. We will use micro-benchmarking tools, such as *fio*, in order to understand systems behavior with fairly simple benchmarks. We will also make use of macro-benchmarks. Regarding this second part, we will base ourselves on a video processing application. The objective of this part is to understand how the storage system can impact video processing applications. Some of the conclusions that we will draw are generalizable to many applications.

In fact, in the context of research projects that one of the authors has pursued at the IRT (Institut de Recherche Technologique) b<>com, in the

Cloud Computing laboratory, we were drawn to studying the impact of storage on the performance of video processing applications. The long-term objective was to propose a video processing service on a Cloud-based platform. This application will be considered as a case study in the next chapter.

8.1.2. Hardware platform

The analysis of the energy consumption of the SSD will be achieved by means of two techniques. The first technique makes it possible to measure the overall energy consumption of a system according to the used storage system. To this end, we have used a Power Distribution and metering Unit (or PDU). A PDU is a device for the distribution of electric power in rack servers and other equipment. A metered PDU allows us to measure the power consumption per electric outlet. Therefore, by means of a PDU, we can measure the overall energy consumption of a given server. The second technique that was used and which allows for a finer measurement, specific to the storage system, is the use of a device that is capable of measuring the energy consumption of storage devices. For this purpose, we have inserted a power sensor on the power cable of the SSD and of the HDD (described in section 8.5). This allows a finer and more accurate measurement than the PDU. These two views are complementary: the overall measurement shows the impact of I/O operations on the overall performance of a system, whereas the finer measurement makes it possible to measure the precise difference between different storage devices in terms of behavior and energy consumption.

8.2. I/O software stack in Linux

Before going into the details of the methodology of performance and power consumption analysis, we are first going to present the I/O software stack in Linux.

Part of this stack has been studied in Chapter 4 (see Figure 4.2). Within the context of the integration of flash memories with a software management layer integrated in a dedicated file system, the I/O software stack is different from the more conventional one used in other Linux systems. In fact, when using a dedicated file system (*FFS*), the latter is interfaced with a generic NAND driver (the MTD layer). On the other hand, in a conventional system, it is the block device layer that is used. The latter contains several sublayers whose features are summarized in this section.

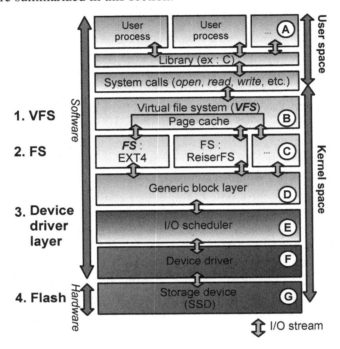

Figure 8.1. *Conventional I/O management software stack in Linux*

Figure 8.1 shows the different layers involved during the processing of I/O requests, specifically requests for accessing files under Linux. We try to summarize the role of each of these layers (the role of the VFS layer as well as that of file systems have already been addressed in the previous chapters, here we will describe only the role of the three new layers) [BOV 05]:

– *The generic block layer*: the generic block layer is to the block device driver what the virtual file system is to the concrete file system. It allows centralizing generic management of block-based devices and abstracting the complexities and specificities of each device (CD-ROM, hard drive, etc.). This layer is also capable of defining structures related to concepts such as the disk, the partition, etc.

– *The I/O scheduler*: the purpose of the I/O scheduler, initially developed for hard drives, is to group the I/O requests and to reorder them to reveal sequentiality and thus increase performance on hard disk drives. Several policies are implemented in the Linux kernel.

– *The device driver*: at this level, requests are transformed into commands that are able to be understood by the storage device.

8.2.1. *The generic block layer*

The generic block layer addresses any request to the block devices, i.e. storage devices (CD-ROM, hard drives, DVD, etc.). Some of the functionality of this layer can be summarized as follows: management of the memory buffers containing the I/O data, implementation of the *zero copy* scheme which makes it possible to avoid having multiple copies of data during I/O transfer (only one copy is maintained and kept in the user space), management of logical volumes (or LVM for *Logical Volume Manager*) and the use of advanced mechanisms such as some options of DMA (*Direct Memory Access*), advanced management policies of disk caches, etc.

The key data structure of this layer is the bio or *block I/O* structure representing the descriptor of an I/O request. This structure contains several fields allowing the user to define a request: the identifier of the drive related to the request, the starting sector number, the number of requested sectors, the memory blocks to be used for the copy of the data of the request (in writes or reads).

Another important structure managed by this layer is the gendisk (*generic disk*) structure, a structure for managing disks in a generic manner. This structure is a means of representing physical disks, such as CD-ROMs, hard disk drives or diskettes, and also virtual disks consisting of several physical disks. This structure contains several fields used to define the disk towards which an I/O request is sent. Among these fields, we can find: the major number and the minor number associated with the disk[1], the name of the disk, a table containing the partitions of the disk, a request queue (see the next section), etc.

---

1 In Linux, a major number and a minor number allow a device to be identified in a unique manner.

On the one hand, the main role of the generic layer is to allocate the `bio` structure and to initialize it with the `bio_alloc()` function, and also to launch the request by means of the `generic_make_request()` function. The first step essentially consists of the definition of fields related to the request: starting sector number, size of the request in sectors, the type of operation (read/write) and a few elements about the device. The second step allows several verifications to be carried out before sending the request to the lower layer, the I/O scheduler. The `generic_make_request()` function verifies, for example, that the number of sectors is not greater than the size of the device. It identifies the correct request queue specific to the device, checks whether the device is a partition or a disk and invokes the I/O scheduler.

8.2.2. *The I/O scheduler*

Initially, I/O schedulers have been designed considering hard disks as storage devices. In fact, with this type of medium, sending individual requests one after the other can result in significant performance degradation. This is mainly due to the mechanical movements that the bad management of these requests can generate. The aim of I/O schedulers was thus to reorder the requests in order to benefit from the performance of sequential access to hard disk drives. Therefore, when a request arrives at the I/O scheduler level, the latter is not immediately satisfied. The kernel attempts to insert it within a sequential flow in order to optimize the overall performance of the access to the disk.

I/O schedulers make use of request queues, each queue holding I/O requests for a given physical disk (block device). The I/O scheduling policy is run for each request queue. Each queue is represented by a `request_queue` structure containing several parameters including the scheduling policy of I/O requests, also called *elevator*, due to the fact that the optimization problem of disk head read/write movements is similar to the optimization problem of the movement of an elevator. The request queue is mainly a doubly linked list containing the request to be satisfied on a given physical block device. A request is represented by one or more structures of the `bio` type.

When a request is sent by the generic block layer, the I/O scheduler receives it and inserts it at the correct location according to the policy under use. The objective is to keep a list of requests, sorted according to the sector number such that sequential access is ensured. The list of requests is usually sorted in order to allow as many requests as possible to be satisfied when the read/write head moves, for example, from the outer track to the inner track. Arrived at the inner track, the direction is changed and requests are satisfied towards the external track. If, during this second pass, a request for the inside track arrives (as other requests to the external tracks also arrive), this request will have to wait until the read/write head reaches the end and comes back in the opposite direction. This can be detrimental.

We will briefly describe in this section four I/O scheduling algorithms existing in the Linux kernel since version 2.6:

– The *Noop* algorithm: this is the simplest algorithm because it consists in doing nothing in particular (*No operation*), i.e. sending the requests originating from the top layer (generic) to the driver in the order of arrival.

– The *CFQ* algorithm for *Complete Fairness Queueing*: the purpose of this algorithm is to ensure fair sharing of the disk bandwidth between the processes launching the I/O requests. Several queues are maintained to store requests for different processes. The algorithm selects a set of requests from each queue and places it inside a queue (the *dispatch* queue) that it reorders according to the sector number. Then, it sends the requests to the device driver.

– The *Deadline* algorithm: the purpose of this algorithm is to overcome starvation problems while trying to optimize performance. This algorithm uses 4 queues, 2 of them containing all of the requests sorted by sector number with a queue per operation type (read, write). By default, the scheduler chooses one direction (usually the one making it possible to satisfy the read request), and sends a set of requests to the *dispatch* queue. The 2 remaining queues contain the same requests classified according to their execution deadline date (separated into read requests and write requests). The deadline is a period of time allocated by the kernel upon the reception of each request, it is initialized to 500 ms for reads and 5 seconds for writes (as a matter of fact, it is more important for users to have their read request satisfied; however, there is usually no need to know when a write request is actually sent to the disk). This deadline value is periodically decremented and when it reaches zero, the request is scheduled in priority with regards to the requests of the 2 sorted queues.

– The *Anticipatory* algorithm: this algorithm is an extension of the previous one and still contains 4 queues, including 2 sorted according to the sector number and 2 according to the deadline. On the other hand, a first difference is the expiration time which is 125 ms for reads and 250 ms for writes. Another major difference is the direction of the movement of the read/write head: one does not always move in one direction until the end. In this algorithm, the read/write head can change direction if the distance to the next sector in the opposite direction is half less significant than in the original direction. This algorithm collects statistics regarding access patterns to queues and aims to anticipate some of the accesses in order to load the data while operating (moving the disk head from an extremity to the other).

Once a request has been chosen by the scheduler, it is sent to the lower layer that of device drivers.

8.2.3. The device driver

At this level, requests are transformed into commands that can be understood by the storage device. It involves the implementation of device drivers allowing reading and writing to disks via DMA transfers (Direct Memory Access) and interrupt management. Device driver developement for disks falls outside of the scope of this book, the interested reader can, nonetheless, access detailed information in specialized books [COR 05, VEN 08, COO 09, LOV 10].

8.3. Context: the Cloud

The work described in this chapter has been achieved as part of a project about storage optimization for an *Infrastructure as a Service* (IaaS) Cloud. An IaaS Cloud is one of the major Cloud Computing services alongside *Platform as a Service* (PaaS) and *Software as a Service* (SaaS). IaaS Cloud aims to provide access to computing resources in a virtualized environment. The resources provided are expressed in terms of hardware, for example number of CPUs, memory and storage space. The customer is, therefore, offered one or more virtual machines (VMs) with specific configurations in terms of allocated hardware resources.

Figure 8.2 shows an overview of our architecture. Clients are allocated one or more VMs, the sharing of hardware resources being managed by a hypervisor running on a host operating system (the one installed on the Cloud servers). VMs access storage devices via the hypervisor and the host operating system. A VM has its own storage space represented by an image file (VMx.img in the figure). Each VM runs its own operating system.

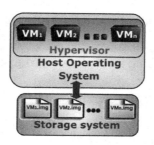

Figure 8.2. *Overall architecture in our Cloud context*

Compared with a conventional computer system, the architecture previously exposed presents an additional layer of complexity due to virtualization. Indeed, in a conventional system, processes directly access files by means of the kernel. In our case, we have an additional virtualization layer. However, the methodology that we shall follow can be applied to both types of systems. We will identify in the following the steps that are not necessary for a system without virtualization[2].

8.4. I/O monitoring tools for performance exploration

In the context of the conducted project, the overall objective was to understand the behavior of the access to the VMs storage systems in order to allocate the appropriate storage devices (magnetic hard disk drives or flash memory). For this purpose, it was essential to build a trace of the I/Os of each VM and thereof to understand the behavior.

As shown in Figure 8.1 and Figure 8.2, an I/O request goes through several layers starting with the operating system of the VM and then through the several layers of the host system: VFS, FS and block layer (generic, scheduler and driver). In order to analyze the I/O behavior of a VM, it is

---

2 Several tools, illustrations and parts of text are taken from Mr. Hamza Ouarnoughi's PhD thesis.

necessary to follow the path of I/O requests down to the storage device. To this end, we have developed monitoring tools (in the same spirit as what has been shown in Chapter 5).

In the context of this study, it was necessary to monitor I/O requests at four different levels:

– Level 1: this first tracing level allows information to be retrieved at the hypervisor level and therefore close to the VM. The goal being to understand the high-level behavior of VMs in terms of I/O.

– Level 2: the second tracing level is that of VFS (as shown in Chapter 5), the objective is to monitor high-level access within the host system. As a matter of fact, several optimizations can be applied at the hypervisor level, modifying the I/O trace arriving to the host system.

– Level 3: at the VFS level, I/O tracing is given in terms of file number and offset within the file. At the lower block level, it is given in terms of block number. We thus need to have access to the information about the mapping between files and blocks. This information is maintained at the file system level, hence the monitoring at this level.

– Level 4: the last level is the block level. Tracing is performed at that level in order to find out which of the I/O requests among all application requests will actually reach the storage device. To this end, block-level tracing is necessary.

In the following sections, we will describe each tracing level and the tools available, enabling I/O monitoring to be performed.

8.4.1. *Level 1: the hypervisor*

This trace level is useful if it is desirable to trace in a virtualized environment, as for instance in the Cloud. In this context, several VMs can be executed and managed by a hypervisor. The hypervisor is the first point of interaction between VMs. The choice of technology to use for tracing is one of the major problems which a developer may be faced with. In an attempt to partially address this problem, we have used *Libvirt*, a library, an API and a *daemon* capable of managing virtualization in a homogeneous manner. *Libvirt* supports multiple hypervisors including KVM/QEMU, VMware, VirtualBox and Xen.

At the level of the I/O virtualization layer and by making use of *Libvirt* APIs, we recover statistics about I/Os, namely about read and write operations executed by each VM in each image. In our case, *Libvirt* allows us to query the virtualization layer on the use of the storage resource in a periodic manner.

Algorithm 8.1. Hypervisor level traces

Data: Hyp=$\{vm_1, vm_2, ..., vm_n\}$
Result: Traces=$\{t_1, t_2, ..., t_n\}$
initialization stats_I/O   /* Access all the virtual machines being
    executed by the hypervisor   */
for *each Virtual Machine* vm_j *in Hyp* **do**
    for *each Virtual Disk* vd_k *attached to* vm_j **do**
      | stats_I/O = Libvirt.stats(vm_j, vd_k) write(t_i) = $\{vm_j, vd_k,$ stats_I/O$\}$
    end for
end for

Algorithm 8.1. shows the function used to retrieve statistics at the hypervisor level. This function is called periodically. It produces a set of trace files as output, one file per hypervisor (or per virtual machine). The tracer goes through all operating VMs in each hypervisor and then the virtual disks attached to each VM. Next, it requests I/O statistics that correspond to each virtual disk. Finally, the traces of each virtual machine are updated. The function produces one trace file per VM with the following format:

```
<vm>;<vd>;<t_stamp>;<#read>; <#size_r>;<#write>;<#size_w>
```

The different fields constituting each trace line are defined as follows:

– `<vm>`: the identifier of the VM performing the I/O operations;

– `<vd>`: the virtual disk on which the I/O operations have been carried out;

– `<t_stamp>`: the time in seconds at which the hypervisor has been queried, the results per time unit represent the statistics accumulated since the last measurement;

– `<#read>`/`<#write>`: the number of read/write requests performed by the VM at the hypervisor level;

– `<#size_r>`/`<#size_w>`: the amount of data read/written in bytes per operation type.

This level of I/O monitoring allows the evaluation of the volume of I/O achieved for each VM. Obviously, this level is insufficient because a VM that is sending application I/O requests does not mean that it is soliciting the storage system and this is due to the fact that there are several existing intermediate cache levels as described in the previous chapters. It is therefore necessary to investigate the I/Os at a lower level.

8.4.2. Level 2: host VFS

The hypervisor retrieves the I/O requests of the different VMs and transfers them to the host system. The latter processes them through the first layer being traversed (as described in the previous chapters), the VFS.

We choose to present two different ways to monitor accesses at the VFS level: (1) by employing the existing *strace* tool or (2) by developing a Linux module as shown in Chapter 5 (VFSMon).

8.4.2.1. Example of the strace tool

strace allows for tracing the system calls executed by a process by specifying its identifier (PID). In our study, we are interested in the specific calls to I/Os on the virtual disks. Therefore, we proceed in two steps: (1) monitor the processes of the VMs and (2) perform a post-processing filtering of the traces obtained in order to keep only those which concern the I/O operations on virtual disks.

a) *Extraction of traces with* strace: In order to monitor each VM, it is necessary to know the PID of the latter (each VM is represented by a process). For this purpose, we have used the API of the library *libvirt* to retrieve this identifier from its name. The PID of each VM is therefore passed as an argument to *strace*.

The system calls of the VM are filtered on the fly to keep only those relating to I/Os. We therefore have monitored the following calls: creat(), open(), read(), write(), lseek() and close(). Concerning on-the-fly filtering, the *-e trace=* option of *strace* allows the user to specify the system calls to be traced.

Timestamping system calls is very important to obtain the sequence of I/O operations at the different trace levels (hypervisor, VFS and block). *strace* allows accuracy at the microsecond granularity. Adding the option *-ttt* displays the time in microseconds before each system call. The following command presents an example of the use of *strace* in our context:

```
$ strace <vm_pid> -ttt -e trace=creat,open,read,write
lseek,close
```

The trace file contains all of the I/O system calls executed by the VM. The format of each line is as follows:

```
<time_stamp>;<i/o_systemcall_with_args>;<return_value>
```

The first field `<time_stamp>` presents the issuing time of the system call in the following form: `seconds.microseconds`. The second field `<i/o_systemcall_with_args>` represents the system call in the form *func* $(arg_1, arg_2, ...)$, where *func* is the system call and $(arg_1, arg_2,...)$ are the values of the arguments of the function. The last field `<return_value>` is the return value of the function (e.g. the number of written/read bytes for primitives `read`/`write`).

It should be noted that a VM does not only manipulate the files of the virtual disks but also other files (e.g. configuration files, binary files, libraries, etc.).

It is therefore necessary to apply another filter so as to keep only accesses to the files of the virtual disks of a VM.

b) *Isolation of traces per virtual disk:* To keep only the I/O operations in virtual disks, we applied a second filtering stage. We used the APIs of the library *libvirt* in order to identify the files representing the virtual disks. Once the virtual disks have been identified, we applied a filter capable of keeping in the trace only system calls relative to virtual disks. This is achieved by retrieving the identifier when initializing the virtual disk and filtering read and write calls with this identifier.

8.4.2.2. *Example of a module with* Jprobe

If it is desirable to track specific functions of the kernel, we can achieve this by developing our own module as shown in Chapter 5. The major difference with what has been previously described in this chapter is the choice of the functions to trace. It turns out that in the context of a system with MTD, the function to trace on the VFS output path will be specific to the FFS. However, within a more general context (for a *ext4* file system for instance, see further below), this will be a different function. For this project, we have traced the following functions, which are among the last being called in the VFS layer:

– for read operations: `generic_file_read_iter`;

– for write operations: `__generic_file_write_iter`.

8.4.3. *Level 3: file system*

At the file system level, we did not develop any specific tracer. On the other hand, monitoring at this level is necessary in order to obtain the link between traces at the VFS level and those at the block level. As a matter of fact, it is at the level of the file system that the mapping between file name and block number is achieved. We must thus extract this mapping in order to establish the link between VFS-level traces and block-level one.

To our knowledge, there is no generic tool allowing us to query the whole set of file systems. Therefore, we have decided to focus on the mostly used file systems, namely EXT2/3/4. For these file systems there is a library, *libext2fs*, providing a mechanism to explore the characteristics of these systems and we have used it.

The original purpose behind the development of *libext2fs* was to allow programs to run in user space to access and manipulate *ext2* system files [TS' 05]. *Libext2fs* proposes an API in C that can be integrated into the monitoring process. To make use of this API while performing I/O operations at the VM level, we need two essential pieces of information:

1) The set of partitions (formatted in EXT2/3/4) in which VM virtual disks are stored.

2) The file identifiers (called *inode* numbers) representing the virtual disks of VMs.

It should be noted that two files stored on two different partitions can have the same *inode* number.

8.4.4. *Level 4: block layer*

Similarly to the VFS level, block-level tracing can be performed in two different ways: (1) using the conventional and very complete *blktrace* tool that is designed to satisfy most requirements or (2) developing a custom tracer using *Jprobes*.

8.4.4.1. *Example of the blktrace tool*

blktrace is a tool that makes it possible to trace a request from the time it is initiated at the level of the generic block layer and down to its processing by the storage device.

Just as the VFS level, the use of *blktrace* for tracing is conducted in two phases. The first step consists of using the tools *blktrace* and *blkparse* to trace I/Os in a physical disk or partition. The second step consists of isolating traces per VM/virtual disk.

a) *Disk or partition trace:* The tool *blktrace* takes a partition or a disk as a parameter. In this context, we assume that the VMs utilize a dedicated disk to store all of the files of the virtual disks (e.g. /dev/sdb). The output produced by *blktrace* cannot be directly employed for analysis. It is necessary to use the tool *blkparse* which takes the output of *blktrace* on input and produces a readable trace, ready to be used for analysis purposes. It should be noted that *blkparse* can also be utilized with *blktrace* on the fly.

The following command shows an example of using *blktrace* in our context, where it is assumed that the virtual disks are stored in a device represented in the system by the file /dev/sdx:

```
root@debian:~# blktrace -d /dev/sdx -o - | blkparse -i -
```

This command is executed by a system thread, in parallel to other trace processes on the other levels (i.e. hypervisor and VFS). The output of this command does not allow distinguishing between the accesses of the various VMs in the virtual disks. This output has to be post-processed in order to isolate the I/O operations of VMs.

b) *Filtering of traces per* VM/*virtual disk:* The program *blktrace* is not capable of tracing the accesses of a particular process but rather accesses in a partition or a particular disk. In our context, it turns out that the traces obtained cannot be used as they are: they have to be processed. The objectives of this post-processing are:

1) Isolating per VM.

2) Establishing a mapping between the block numbers and virtual disks for the traces of a same VM.

3) Filtering of I/O operations to keep only read and write operations.

4) Filtering of I/O scheduler operations in order to keep the requests actually processed by the storage device.

Figure 8.3 shows an example of tracing a VM executing I/O operations, obtained by means of *blktrace*.

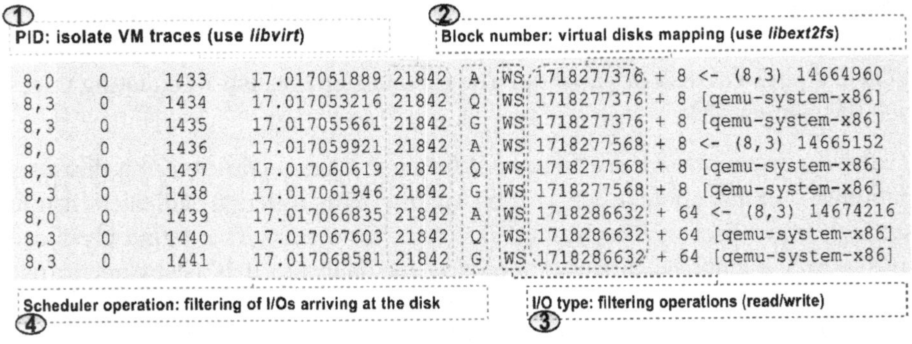

① PID: isolate VM traces (use *libvirt*)   ② Block number: virtual disks mapping (use *libext2fs*)

```
8,0    0    1433    17.017051889 21842   A   WS 1718277376 + 8 <- (8,3) 14664960
8,3    0    1434    17.017053216 21842   Q   WS 1718277376 + 8 [qemu-system-x86]
8,3    0    1435    17.017055661 21842   G   WS 1718277376 + 8 [qemu-system-x86]
8,0    0    1436    17.017059921 21842   A   WS 1718277568 + 8 <- (8,3) 14665152
8,3    0    1437    17.017060619 21842   Q   WS 1718277568 + 8 [qemu-system-x86]
8,3    0    1438    17.017061946 21842   G   WS 1718277568 + 8 [qemu-system-x86]
8,0    0    1439    17.017066835 21842   A   WS 1718286632 + 64 <- (8,3) 14674216
8,3    0    1440    17.017067603 21842   Q   WS 1718286632 + 64 [qemu-system-x86]
8,3    0    1441    17.017068581 21842   G   WS 1718286632 + 64 [qemu-system-x86]
```

④ Scheduler operation: filtering of I/Os arriving at the disk   ③ I/O type: filtering operations (read/write)

Figure 8.3. *Parameters utilized for post-processing*

Figure 8.3 shows the usage of the different trace fields obtained in the post-processing procedure. The PID of the process executing the I/O requests allows us to isolate the traces to distinguish VMs. This is the same PID employed to monitor the VFS level. The operation type field allows us to filter so as to keep only read/write requests. The field indicating the operations carried out at the I/O requests queue level is used to keep only completed I/O requests.

8.4.4.2. *Example of a module with* Jprobe

The second way to trace block-level accesses is to develop a custom module. Thus, we have to first choose the functions to monitor. We have decided to trace the `generic_make_request()` function, which, as it has previously been seen, is called from the generic block layer (declared in the Linux kernel sources in `Linux/include/Linux/blkdev.h` and implemented in `Linux/block/blk-core.c`).

The prototype of the function `generic_make_request()` is defined as follows:

```
1
2 /* bio: structure describing the I/O operation */.
3 blk_qc_t generic_make_request(struct bio *bio)
```

This function is used to perform I/O requests described in the `bio` structure. In fact, this structure allows us to extract all necessary information on I/O operations being executed. Here follows the `bio` structure with just the fields being used in our tracer:

```
1 struct bio {
2 struct block_device *bi_bdev; /* Storage device */
3 unsigned long        bi_rw; /* Type of I/O operation
     */
4 struct bvec_iter     bi_iter; /* Block number, number
5                    of blocks and inode number */
6 };
```

The parameter `bi_bdev` represents the storage device (*block device*) or the partition in which the I/O operation is executed. This parameter allows us to distinguish the different disks (partitions) and thus to be able to trace several devices at a time. The parameter `bi_rw` gives the type of I/O operation and can have READ or WRITE as value. `bi_iter` enables us to know the number of the starting block of the I/O request, the size of the request (in blocks) and the *inode* of the file used in the I/O operation being performed, and therefore to establish a connection with the high layers of the tracer.

Here is an example of tracing in VFS and block levels:

```
1 436.809786717;R;2051751936;4096;VFS;qemu-system-x86
     ;875
2 436.810379949;R;1902489600;4096;VFS;qemu-system-x86
     ;8751
3 436.810398806;R;619393760;16384;BLK;qemu-system-x86
     ;8751
4 436.811018927;R;1805774848;4096;VFS;qemu-system-x86
     ;8751
5 436.811059225;R;616685128;8192;BLK;qemu-system-x86
     ;8751
```

The output of the tracer is of this form:

```
<time_stamp>;<i/o_type>;<blk_nbr>;<blk_size>;<level>
;<proc_name>;<PiD>
```

with:

- <time_stamp>: start time of the request;

- <i/o_type>: operation type (read or write);

- <blk_nbr>: block number;

- <blk_size>: operation size;

- <level>: level (VFS or block);

- <proc_name>: name of the process that launched the operation;

- <PiD>: identifier of the process.

This tracer, developed in the context of a project of the IRT b<>com and whose description can be found in [OUA 14], can be downloaded at *https://github.com/b-com/iotracer*.

8.5. Performance and energy consumption analysis

As discussed in the introduction, we make use of two different methods in our measurements: (1) using a PDU making it possible to get the measurement of the overall power consumption of the system and (2) using a sensor allowing the measurement of storage devices power consumption. Two storage devices have been tested: an SSD and a HDD in order to highlight the existing difference in behavior. In this section, we will describe these two measurement methods for our experiments and whose results are exposed in the following chapter.

8.5.1. *Measurement of the overall system (PDU)*

We have measured the overall electric power of a system using the Raritan PX3 5190R PDU. Figure 8.4 illustrates the measurement platform by making use of a PDU.

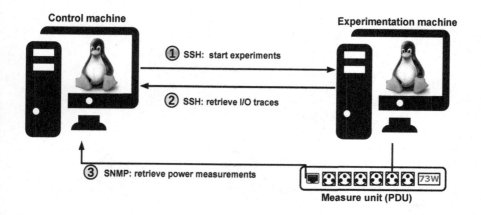

Figure 8.4. *Measurement platform with a PDU*

As shown in Figure 8.4, we have employed three devices to achieve our power measurements:

1) *A control machine*: from which experiments are launched and in which I/O trace analyses are carried out.

2) *An experimentation machine*: which contains the storage devices (HDD and SSD) that host VMs and execute their I/O workloads.

3) *A measurement unit (PDU)*: which is responsible for measuring the electric power of the experimentation machine during the execution of the different tests.

Figure 8.4 also gives an overview of the three main stages of running a test:

– the start of experiments using the SSH communication protocol in order to minimize the intrusivity of the control of experimentations in the tests being carried out;

– the extraction of I/O traces resulting from the execution of the experiments;

– the extraction of the values of electrical power measured while executing the test. This step is performed in parallel with step (2) and the protocol used to send commands to the PDU is SNMP.

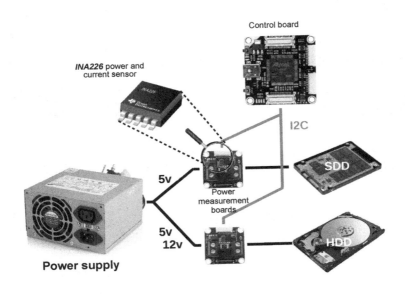

Figure 8.5. *Measurement platform with power sensor*

8.5.2. *Measurement of the storage system (sensors)*

The second measurement method consists of isolating the energy consumption specifically related to the storage device. For this purpose, we have developed an environment for the measurement of energy consumption as described in Figure 8.5. This environment is composed of two types of boards:

1) A board containing the power sensor: the sensor employed is the *INA226* from *TI*[3] capable of measuring the current, voltage and power and transmit them via a I2C communication bus. This sensor is shipped on a board[4] which makes it possible to insert it more easily between the power supply and storage devices (see Figure 8.5). The SSD is powered by a 5 V voltage whereas the hard disk drive uses two power supplies: 5 V and 12 V (for the motor). Therefore, we have used two boards for the HDD and one for the SSD (see

---

3 http://www.ti.com/lit/ds/symlink/ina226.pdf
4 http://www.tinkerforge.com/en/doc/Hardware/Bricklets/Voltage_Current.html

Figure 8.6). The sampling rate is of the order of 1 KHz. At 1 KHz, one can count a power measure every millisecond, which is very accurate for events that we intend to measure, in this case I/O operations;

2) A control board: in order to recover the power measurements from the board containing the sensor, we have employed a control board[5]. This board contains a network interface making it possible to be accessed by a remote machine in order to harvest the measures (as well as PDU measures). This board communicates with the sensor board via an I2C interface.

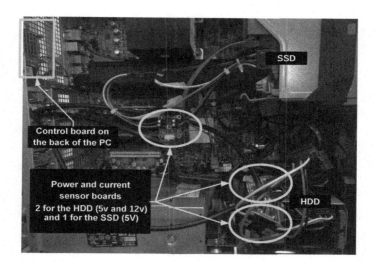

Figure 8.6. *Picture of the platform used*

8.6. Conclusion

In this chapter, we have described all the elements enabling energy consumption measurements to be performed on storage devices.

From the hardware perspective, we make use of two measurement methods: an overall measurement of the system with a PDU and a finer measure with sensors. From the system perspective, we have described the

5 https://www.tinkerforge.com/en/shop/bricks/master-brick.html

overall functioning of the I/O operations on storage devices and have described the different tracers allowing us to understand the I/O workflow. From the application perspective, we mainly focus on the behavior of energy consumption with micro-benchmarks. Subsequently, we study the behavior with a real application, in this case, a video processing application.

In the following chapter, we will describe the results obtained with this measurement methodology, this will allow us to verify to what extent storage can have an impact on energy consumption and we will pinpoint the differences between SSD and HDD.

9

Performance and Power Consumption of SSD Based Systems: Experimental Results

In this chapter, we aim to describe the measurements that have been performed on real storage systems in order to understand the performance and energy behavior of SSDs and HDDs. To this end, we rely on the methodology and the tools described in the previous chapter.

This chapter is structured around three sections:

1) The first section describes the impact of I/Os on the energy consumption of computer systems.

2) The second section concerns performance and energy consumption of computer systems according to the storage device and I/O patterns. It provides a macroscopic view depending on the storage system being used and file access patterns.

3) The third section presents a more microscopic view and focuses on the power consumption of the storage device. We will describe the results of experimental measurements for multiple disks (HDD and SSD) in an attempt to identify their behavior.

9.1. Introduction

Through this chapter, we will strive to understand the impact of I/Os in a computer system, more specifically concerning accesses to flash memory in an SSD (as previously described, equipped with an FTL). We also carry out a comparison with hard disk drives.

As indicated in the previous chapter, we will study the power consumption mainly by means of state-of-the-art micro-benchmarks. We will also perform a few experiments with a video application in order to observe the impact of storage on a real application. In the first case, we will consider the overall energy consumption of the system measured with a PDU and the specific consumption of the disks that were used: HDD and SSD. In the second, we will focus on the behavior of the storage system.

9.2. Impact of I/Os on performance and energy consumption

Before beginning to describe the experimental aspects, it proves interesting to take a look at the behavior of I/Os and the impact of processing I/Os on the energy consumption of a computer system.

As a matter of fact, it is very important to understand that when accessing a file through read or write operations, the storage device is not the only element impacted. It turns out that during the whole duration of the access to the storage system, the CPU and the memory continue to consume energy, a portion of energy that is even much more significant than that consumed by the disk during I/O processing. When the system has several tasks to be performed and when a given process intends to perform a read or write to a file, since this operation takes time, the kernel changes the state of this process from active to blocked (pending the completion of the I/O request). The kernel can then start executing a new process while the I/O request is being processed. In this case, the processor is being solicited but it is actually processing another task. However, if the system does not have a particular task to run, the CPU goes on consuming without achieving any useful work. Policies to reduce the frequency can be effectively implemented (DVFS for *Dynamic Voltage and Frequency Scaling*), which allows power consumption to be reduced. Nevertheless, this energy consumption remains more significant than that of the storage device.

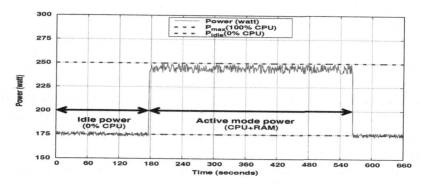

Figure 9.1. *Behavior of the system power consumed without storage system. For a color version of this figure, see www.iste.co.uk/boukhobza/flash.zip*

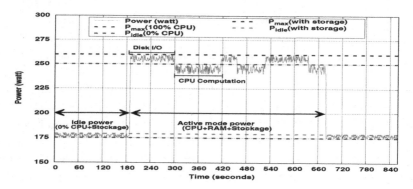

Figure 9.2. *Behavior of the system power consumed with storage system. For a color version of this figure, see www.iste.co.uk/boukhobza/flash.zip*

Figure 9.1 and 9.2[1] illustrate what has been described in the previous paragraphs. In the first figure, the behavior of a system that does not access to storage is shown. In these figures, the measures represent the power consumption of the entire system (which can be measured with a PDU). It can be seen that the overall system power consumption when the processor is not in use is 175 Watts. When the processor reaches a 100% utilization, the power consumption increases up to 250 Watts. It should not be forgotten that the energy consumption is the integral of the power and therefore the total surface

---

1 Note that these figures are not the result of measurements and that the given curves are indicative.

below the curve power line. The larger the amplitude of the power, the higher the system energy consumption. Furthermore, the longer the system runs with the same power consumption, the more significant its energy consumption.

Figure 9.2 illustrates the behavior of the same system when we consider the storage device or more specifically the I/O accesses to the files on the storage device. The chart begins with a phase during which the system is idle and thus the power consumption is equivalent to that of a system without workload (175 Watts). Subsequently, the processor becomes active and will initiate I/O operations. We note that the energy consumption increases up to a maximum approximately 260 Watts. This is the active consumption of the CPU and RAM in addition to the storage device. It turns out that during the processing of I/Os, the 3 components consume power. After this I/O phase, we note that the power decreases with a maximum of 250 Watts. This represents the consumption when the disk is idle. Thus, the activity of I/Os impacts on the energy consumption of the system. It should be noted that the disk never consumes energy alone when dealing with I/O requests, but the CPU and the memory are also triggered. As previously mentioned, energy consumption is linked to the value of the power and to the duration during which this power is dissipated. Consequently, I/O operations increase energy consumption twofold: (1) by increasing the amplitude of the total power of the system because the storage device is solicited, (2) by significantly increasing the duration during which the whole (CPU, RAM, storage) is operating to satisfy an I/O request, because I/Os are known to be among the most time-consuming operations. As such, SSDs allow a double advantage compared with HDDs: (1) they consume less power, and (2) they are faster and therefore save time and as a consequence save energy consumption, CPU and RAM. It is especially this second point that enables consequent energy benefits[2].

9.3. A macroscopic view of performance and power consumption of storage systems

In this section, we compare the performance of servers using traditional storage systems (HDD) or flash-based disks (SSD) using three metrics: the

---

2 This is true for a benchmark executing a single *thread* as tested in this book. Regarding computation intensive applications, this may not be true as shown in the section addressing the video application.

data transfer rate, power and energy. This is achieved by making use of the tool *fio*[3]. The configuration of the tools employed as well as the hardware utilized is described in the next section.

9.3.1. *Hardware/software configuration for experiments*

fio, an acronym for *flexible I/O tester*, is a benchmark for file systems developed by *Jens Axboe*, maintainer and developer of the generic block layer of the I/O software stack for Linux and developer of *blktrace*. *fio* was initially developed to test and check for changes in the Linux I/O software stack. The diversity of the parameters provided by *fio* makes it possible to build a variety of I/O workload scenarios. *fio* supports approximately 20 I/O mechanisms (e.g. *libio*, *sync*, *mtd*, *net*, etc.). The conventional method of using *fio* is to build a file describing one or more benchmarking scenarios to apply.

The program *fio* was launched to make sequential and random reads and writes. We did vary the sizes of requests between 2 KB and 1 MB by powers of 2 such that to test a wide range of request sizes. We have tried to reduce as much as possible the effect of the page cache in order to focus on the performance and throughput of the storage device (*IO engine*[4]: *sync (read(), write(), fseek()), Buffered: 0, Direct: 1*). We have created files with which the measurements have been carried out. Moreover, the page cache effect has been verified for each test with our trace tools described in the previous chapter. In fact, we have made sure that all application I/O requests are performed onto the storage media. The file system utilized is ext4.

In our tests, we have used a workstation with an Intel(R) Xeon E5-1620 3.70GHz CPU, with 16 GB of RAM. With regard to disks, we used in this section a HDD of 1 TB with a 64 MB cache Seagate model ST1000DM003, and a disk from Samsung SSD 840 PRO model with 128 GB containing a cache of 256 MB. According to the specifications of the manufacturer, the HDD consumes about 5.9 W when operating and approximately 3.36 W when idle. For its part, the SSD consumes between 0.054 W and 0.349 W when idle according to the activation of the power management mode.

---

3 All of the measurements in this section have been conducted by Hamza Ouarnoughi.
4 Several I/O engines can be used with *fio*.

9.3.2. *Measurement results*

We recall that the objective of this section is to identify the impact of the overall storage performance and power consumption of a computer system.

9.3.2.1. *HDD throughput*

Figure 9.3 shows the evolution of the throughput on a HDD by varying the request size from 2 KB to 1 MB, and this for four different I/O workloads: two types of operations (read and write) and two access patterns (sequential and random).

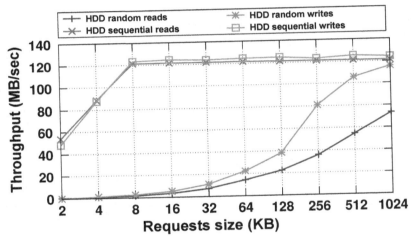

Figure 9.3. *HDD I/O throughput according to the size of requests*

We can observe several phenomena from this curve:

– performance in sequential access is better than with random access and that regardless of the operation being carried out (read or write). In fact, random I/Os generate more movements of the read/write head and larger disk rotation which generates more latencies;

– performance with sequential read and writes is similar. We observe very little difference compared with the type of operation achieved in the case of a sequential pattern. In fact, the generated mechanical movements are the same and even if the magnetic process of reading and writing is different, it generates only very little difference from the performance point of view;

– performance in sequential access for small request sizes is worse than large sizes. It turns out that, the smaller the sizes of the requests are, the more CPU latencies are generated due to the management by the kernel of these I/O requests. Starting from a size of 8 KB and more, these management latencies are much less significant compared with the mechanical latencies caused by the hard disk drive when reading and writing data. Regarding this disk, performances are thus stabilized starting with request sizes of 8 KB;

– with random access, performance improves when increasing the request size. This is due to the improvement of sequentiality. Furthermore, despite the fact that the access pattern be random, with increasingly larger request sizes the access becomes sequential within an I/O request itself thus reducing mechanical movements for a given data volume;

– performance in random access for writes is better than that for reads. This observation is not valid for all of HDDs but strongly depends on the properties and mechanisms implemented by these. This difference is mainly due to the use of buffer memory inside HDD controllers. The latter apply write back mechanisms thus buffering the writes before sending them to magnetic media. This makes it possible to reorder (sequentialize) I/Os and as a result to minimize latencies due to mechanical movements. In addition, the acknowledgment is sent by the disk after writing to the disk cache. This behavior is, however, bounded by the performance in sequential writes.

9.3.2.2. *SSD throughput*

Figure 9.4 shows the evolution of throughput on an SSD by varying the size of the request from 2 KB to 1 MB, and this for four different I/O workloads: two types of operations (read and write and two access patterns (sequential and random). We can observe several phenomena from this curve (we should note that this disk presents atypical performance in random access as we will see further):

– all of the curves follow the same tendency, i.e. the larger the size of the request is, the better the performance. As explained for HDDs, this is mainly due to the latency of the kernel when processing I/O requests. This latency is more visible because the performance of SSDs is better than that of HDDs;

– random reads are performing worse than sequential reads. This behavior is unusual because generally *performance with random and sequential reads are similar* as shown in the curve in Figure 9.5 taken from [PAR 11]. In this study, several SSDs were tested. The unusual performance of the disk that we have tested for this book has been reported in other studies. Nonetheless, this disk is very popular and it is interesting to note that the performance model of SSDs is not always homogeneous;

– sequential and random writes present similar performances. This behavior is not usual. Comparing with the curves taken from [PAR 11]. We note that generally *random writes are performing much worse than sequential writes*. Here again, the disk that we have tested proves that it is difficult to characterize the performance of SSDs without prior testing them. The implementation of FTLs and SSD buffers is very variable from one SSD to another and can lead to quite significant differences of behavior.

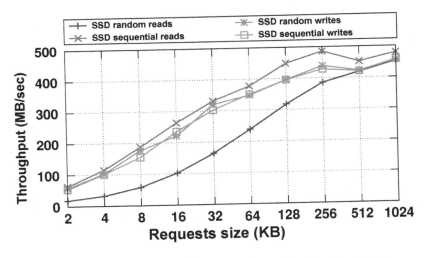

Figure 9.4. *SSD I/O throughput according to the size of requests*

Several SSDs have been studied in [PAR 11]. As shown in Figure 9.5, for the majority of tested SSDs, performance with random reads behaves in the same way as that of sequential reads for a given disk. On the other hand, it can be observed that there is a wide disparity between disks (up to an order of magnitude for the disks tested in this study).

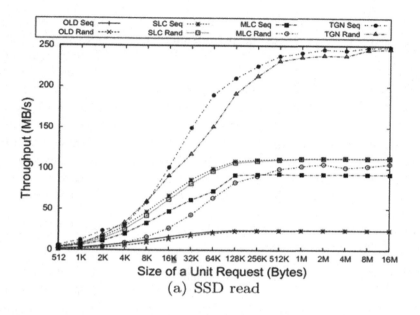

(a) SSD read

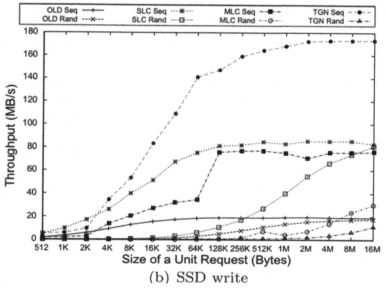

(b) SSD write

Figure 9.5. *Read and write throughputs of several different SSDs [PAR 11]*

The performance curve of writes outcome from this study also shows two important phenomena. First, the performance of sequential writes is much better than for random writes for a given disk. This is a fairly widespread behavior but as it has been seen for the disk we have tested ourselves, this is not always the case. The second phenomenon that can be observed is the disparity of the performance between different existing disks. Indeed, we can note more than one order of magnitude between the performance of the various disks. Moreover, variations in the performance of writes according to access patterns are very different from one disk to another. As an example, the SSD showing the best sequential write performance is the one that has shown the worse performance in random writes (SSD denoted TGN in the curve).

Several differences can be noted between the behavior of HDDs and SSDs from the performance perspective.

First of all, we can note the performance gap between HDDs and SSDs, which can be very significant. For the disks that we have tested, this is of the order of a factor of 5 for large requests sizes and increases up to a factor greater than 30 for small requests.

While the access pattern is a prominent and common criterion for all HDDs, it would seem that it is less significant for SSDs. The discriminating criterion rather seems to be the type of operation: performance is generally higher for read operation than write ones. For reads, performance of sequential and random access is generally similar; for writes, random accesses are generally performing much worse than sequential accesses, because of the latencies of the address translation mechanism and the frequent activation of the garbage collector. That said, some SSDs implement mechanisms that make it possible to improve the performance of random writes.

9.3.2.3. *Power consumption of a system with HDD*

By measuring the power of the server while the latter performs I/O operations, we can make the following observations from Figure 9.6 (it should be noted that what is presented in the curve is the power of the whole server for 1 MB of data):

– the power measure is more significant during sequential accesses, namely for reads and writes. Despite seeming counter-intuitive this finding is definitively not so. It turns out that the more significant the throughput is, the more the CPU will work to process requests, therefore the instantaneous power will be higher if the throughput is higher. This is due to the fact that the power of the CPU is much more significant than that of disks. Consequently, even if the disk shows a higher consumption due to random accesses, it remains hidden because of the CPU consumption;

– a difference in power can be observed between reads and writes. This difference is mainly due to the mechanism of the disk cache responsible for speeding up write operations.

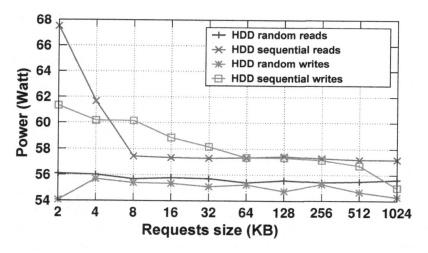

Figure 9.6. *Power consumption of I/O operations in a system with HDD according to the request size. The power is given for 1 MB of data*

9.3.2.4. *Power consumption of a system with SSD*

Concerning the power of servers with SSDs (see Figure 9.7), it can be seen that there is no significant difference between random and sequential accesses, nor between reads and writes. However, as the sizes of requests increase, the power decreases because the CPU processes fewer requests.

The major differences with regard to power between SSDs and HDDs are: (1) the power consumption values: on average, the HDD has a power 3 times higher than that of the SSD and (2) the decrease in power according to the size of the request is smoother in the case of an SSD. We find a symmetric behavior for throughputs.

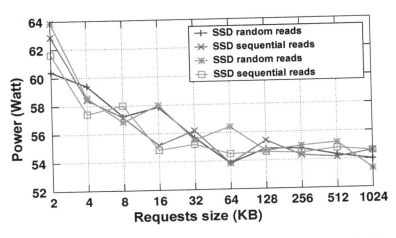

Figure 9.7. *Power consumption of I/O operations in a system with SSD according to the request size. The power is given for 1 MB of data*

9.3.2.5. *Energy consumption of a system with HDD*

As previously discussed, energy consumption varies according to the power and the time during which this power is dissipated. Therefore, the more significant the power amplitude is, the higher the energy consumption. Furthermore, as the execution time increases, the power consumption also increases.

We observe in Figure 9.8 that unsurprisingly, the server consumes a lot more during random accesses, especially with small requests, mainly because of the very bad performance in this case. Although the gap in terms of power is of the order of a few watts, this has a very significant impact due to the very slow response time. With 4 KB requests, a difference of a factor of 100 is observed. Here again, we note a behavior specific to the access pattern with little difference according to the type of operation. The curve of sequential accesses being flattened because of the scale, we reproduce an enlargement thereof in Figure 9.9.

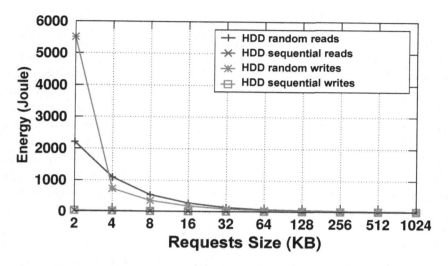

Figure 9.8. *Energy consumed according to the request size in a system with HDD*

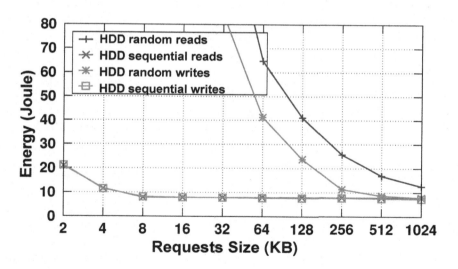

Figure 9.9. *Enlargement of the power consumption figure for sequential accesses in a system with HDD*

9.3.2.6. *Energy consumption of a system with SSD*

With regard to the energy consumption of SSDs (see Figure 9.10), we note an isolated curve in particular, the one of random reads. As previously described, the performance of random reads is especially bad for the SSD that we have tested, and it is for this reason that such significant energy consumption can be observed. As a matter of fact, since the I/O requests last for a long time, the energy consumption is very high. Energy consumption with other access patterns and types of operation are similar. We also observe that the larger the request size is, the greater the decrease in energy consumption. This is due to the convergence of two properties: with increasing sizes, throughput becomes larger and the CPU processes less requests.

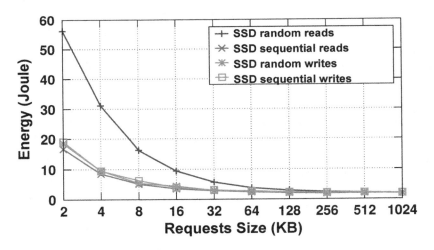

Figure 9.10. *Energy consumed according to the request size in a system with SSD*

9.3.2.7. *Conclusion*

In this section, we have noted several observations concerning the overall behavior of a system with SSD or HDD. First of all, we have observed a different behavior in performance. SSDs have better performance than HDDs and do not exhibit the same performance model. Despite that a tendency appears for SSDs, the latter are less homogeneous than HDDs. With regard to energy consumption, SSDs consume much less than HDDs because the power amplitude is reduced, and in addition SSDs are faster, which allows them to complete I/O processing faster than HDDs.

9.4. A microscopic view of performance and power consumption of storage systems

In this section, we focus on the microscopic behavior of storage devices by studying their energetic properties through measures using sensors as described in the previous chapter (see Figure 8.6).

9.4.1. *The use of micro-benchmarks*

In this section, we will compare the energy consumption of conventional storage systems (HDD) with flash-based ones (SSD). We achieve it by making use of the program *fio*[5]. The configuration of the tools employed as well as that of the hardware experimented is described in the next section.

9.4.1.1. *Hardware/software configuration for experiments*

The utility *fio* has been used for the first part of this section, with the same configuration as described in the previous section.

In this section and with regard to the disks, we have used a hard disk of 1 TB from Hitachi model STHCS721010KLA330 with a 32 MB cache and a 256 GB Samsung SSD model 850 PRO with a 512 MB cache. According to the specification of the manufacturer, the HDD consumes about 5.9 W when operating and approximately 3.36 W when idle. For its part, the SSD consumes about 0.4 W when idle and approximately 3 W when operating.

9.4.1.2. *Measurement results*

The section that follows aims to discuss the results of micro-benchmarks applied to various storage systems with power and energy measures of the storage devices by means of sensors.

---

5 All of the measurements achieved for this section have been carried out by Jean-Emile Dartois, engineer at the IRT b<>com.

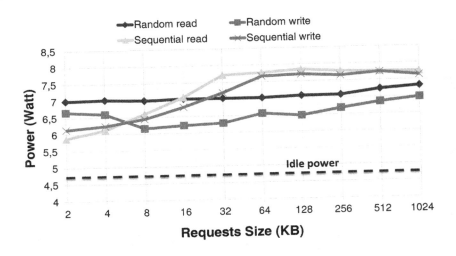

Figure 9.11. *HDD power consumption according to request size*

9.4.1.2.1. HDD power consumption

Figure 9.11 shows that the power of the disk during the various operations is fairly stable, varying between 6 to 8 Watts. Concerning sequential accesses, we note that the power in the device increases with the increase of the request size until reaching a size of 32 KB (for reads) and 64 KB (for writes) where the power stabilizes. We can observe that this corresponds to the stabilization of the throughput at its maximal value (see Figure 9.12). In fact, the more significant the throughput, the higher the consumption of the controller (electronic parts) will be to process higher numbers of requests per unit of time.

The power consumption of random I/Os is fairly stable showing a small tendency to increase. The explanation for the increase of power is the same as previously, this is due to the increase in throughput and thus in the processing of requests by the controller.

We can also note the best behavior of write operations when compared with reads. This is due to the write-back mechanism used in the disk cache. This mechanism enables reordering accesses to sequentialize them when possible. This results in improving the performance and reducing mechanical latencies.

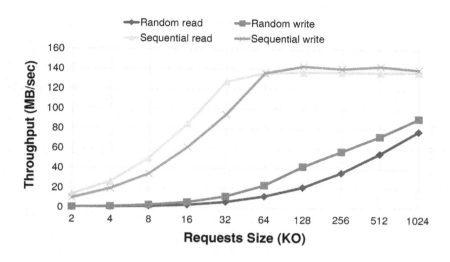

Figure 9.12. *Throughput of I/O operations in a HDD*

9.4.1.2.2. SSD power consumption

With regard to the power of I/O operations on SSD. We note in Figure 9.13 that power increases with the size of the request. More accurately, the power increases with the throughput of the SSD and the more significant the throughput is (see Figure 9.14), the higher the SSD consumption, until stabilizing to approximately 3 Watts.

We can also observe that the smallest power registered is that of random read accesses. However, the difference in performance is higher (more visible) than the power difference. This is explained by the fact that mechanisms internal to the SSD (the FTL) work more, but nevertheless process fewer requests. As a result, this yields a reduced throughput but a power almost as high as with the other access patterns.

Another interesting observation is the difference between the values of energy consumption between the HDD and the SSD. The power of the HDD (∼7 Watts) when processing requests is 3 times greater than that of the SSD (∼2.5 Watts). We can also note the difference in throughputs, which is the same as for the previous experiments: about 140 MB/s for the HDD and 500 MB/s for the SSD.

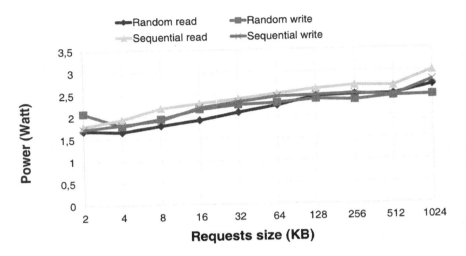

Figure 9.13. *SSD power according to the request size*

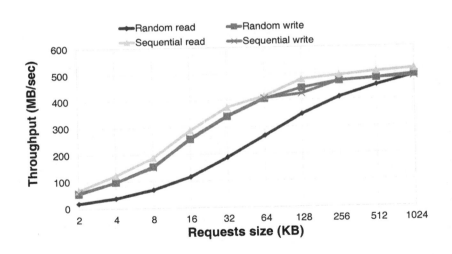

Figure 9.14. *Throughput of I/O operations in SSD*

9.4.1.2.3. HDD energy consumption

With respect to energy consumption (see Figure 9.15), we can see a very large difference between random and sequential I/Os, especially for small request sizes. In fact, because of the slow speed of random accesses, even if the power is not very different from that of sequential accesses, the difference in energy will be very significant. Again, random writes are performing better than random reads because the disk uses write back in its cache. The latter has a size of 32 MB making it possible to buffer and optimize accesses.

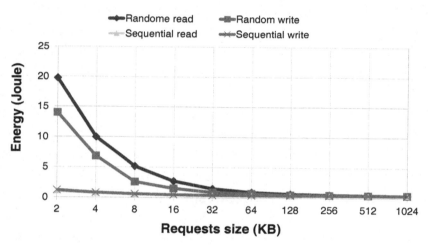

Figure 9.15. *Energy consumed by HDD according to request size*

9.4.1.2.4. SSD energy

As expected, when observing the energy consumption of the SSD (see Figure 9.16), we can observe a very high energy consumption for random reads. Again, this behavior is very specific to the disk used in the test but is symptomatic of a difficulty of generalization regarding the performance and consumption of SSDs according to the type of operation.

As shown in Figure 9.17 according to [PAR 11] and unlike the SSD disk that has been tested:

– the energy consumption of sequential I/O accesses is far better than that of random accesses;

– the least efficient access pattern from the energy point of view is with random writes. This is explained by the latencies introduced by the address translation mechanism as well as the garbage collector of the SSD;

– the energy consumption of the random read I/O accesses is generally very good, except for some SSDs as seen in the experiments that we have conducted.

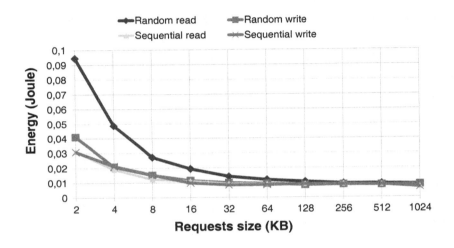

Figure 9.16. *Energy consumed by SSD according to request size*

9.4.2. *Video application case study*

In the context of a research project regarding video processing in the Cloud, we have come to study the difference of the energy consumption between a system with SSD and another with HDD. The objective of the study was to investigate video processing situations that would benefit from a performance of an SSD-based storage system.

In this section, only one single video processing example will be studied.

A video is usually encapsulated in a container (e.g. AVI, mov, ogg, vob, flv). This container comprises a video stream, an audio stream, metadata, sub-captions, chapter information elements, etc.

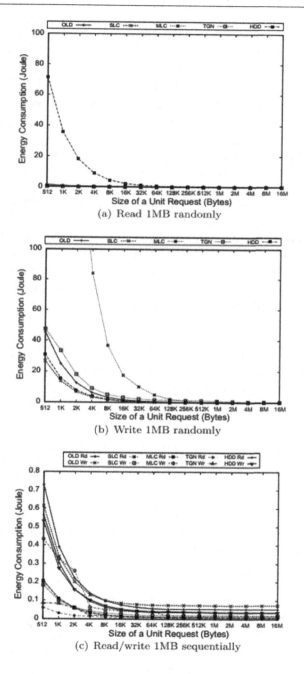

Figure 9.17. *Energy consumed by I/O operations according to request size for various SSDs [PAR 11]*

A video stream can be compressed according to various formats by means of *codecs* (portmanteau word to designate encoder/decoder). For example, the very conventional format H264 or MPEG-4 AVC (*Advanced Video Coding*) and the HEVC H265 (*Highly Efficient Video coding*) supposed to succeed to H264 in the future. In order to compress a video, several parameters have to be specified including the resolution and the bitrate.

We have used the *ffmpeg* framework for our video processing. The program *ffmpeg* is a suite of free software dedicated to audio and video processing.

In our case, during the transcoding of an H264 video with *ffmpeg*, it is possible to vary a parameter called *PRESET*. This parameter has a direct impact on the quality of the compression as well as on the file size. Using a *PRESET* with a value *slow* for a file size given on output, the quality will be better than with a *PRESET* set to *fast*. On the other hand, the transcoding will take longer. If the file size is not defined, a slower *PRESET* will give better compressed files but will take longer.

In this section, we vary the value of the *PRESET* as well as the storage device in order to assess the impact this can yield on the energy consumption. It should be noted that video processing applications are among the most consuming in terms of computing power.

9.4.2.1. *Experimental configuration*

In the tests that we have carried out, we have utilized the server with an Intel(R) Xeon(R) CPU E5-1620 v2 @ 3.70GHz processor, 16 GB of RAM memory and two disks: a Barracuda HDD of 1 TB 7200 RPM and an SSD Samsung SSD 850 Pro of 250 GB.

We have used version 3.1.1 of *ffmpeg* and tracers that we have developed. The test were performed with an HD video of about 700 MB in H264 format. The video has been downloaded from https://mango.blender.org/download/. The bandwidth of the source video is 7862 kb/s, in *ultrafast*, the bandwidth of the encoded video is 101 kb/s and with *slow*, 29 kb/s. We have employed two configurations of the *PRESET*: *slow* and *ultrafast*.

For each of the experiments, our tracers were launched in order to observe the I/O behavior. The power consumption of the two storage devices has been measured.

9.4.2.2. *Results with the* PRESET *set to* ultrafast

Figures 9.18 and 9.19 show three different metrics for the transcoding on hard disk drive: the power consumption of the disk, the amount of data written at the VFS and block levels and the volume of data read. The volumes of data shown on the curves are cumulative.

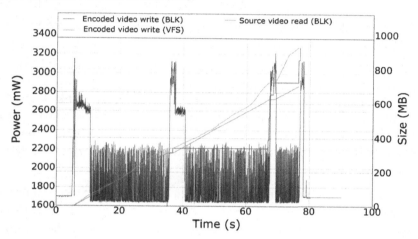

Figure 9.18. *Power consumption and volume of I/Os during the transcoding process using* PRESET *set to* ultrafast *for the 5 V power supply in a HDD. For a color version of this figure, see www.iste.co.uk/boukhobza/flash.zip*

The HDD being powered by 5 V and 12 V voltages, we have decided to show the power for each case (unlike the previous sections where we have shown the cumulative power). HDDs make use of the 12 V voltage for mechanical movement (motor and arm) and the 5 V voltage for the electronic parts and the read and write heads.

We can make several observations from the curves of Figures 9.18 and 9.19:

– *ffmpeg* performs reads and writes during the whole transcoding process. This can be seen in the green and light blue curves respectively, showing reads at the block level and writes at the VFS level captured with our tracer. At the application level, reading and writing data are achieved on a continuous basis during all the test;

– postponing writes at the block level (therefore lower than VFS and closer to the hardware) is achieved by bursts. It turns out that this behavior, captured by the block-level tracer, allows us to observe that the page cache buffers the writes before sending them to the storage device. The volume of memory used in this buffer depends on the available system memory. Periodically, a kernel thread flushes this buffer every 25–30 seconds. In addition, we note a last flushing at the end of the test once the transcoding is complete. This flush operation has not been initiated by the kernel but by *ffmpeg*. In fact, when video transcoding is complete, *ffmpeg* synchronizes the file on disk by means of a flush operation;

– using the tracer that we developed, we have also observed that I/O requests from applications have been grouped in order to optimize I/Os to the storage device. We have noted a difference between VFS and block levels. It turns out that at the VFS level, the tracer has identified that 65.61% of the requests (about 60000) had a size smaller than 844 bytes. Then, at the block level, the number of requests has decreased to 1518 with a size of 688 KB for 91% of them;

– during the flush operation of the data in the page cache, we can observe power peaks in the curves of the two different voltages. These peaks are particularly visible for the 5 V voltage, the one responsible for the read/write heads as well as for the controller.

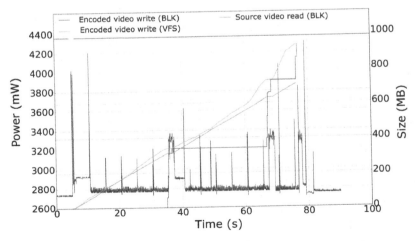

Figure 9.19. *Power consumption and volume of I/Os during the transcoding process using* PRESET *set to* ultrafast *for the 12 V power supply in a HDD. For a color version of this figure, see www.iste.co.uk/boukhobza/flash.zip*

Figure 9.20 shows the total power consumption of the HDD during transcoding. In this figure, the last three consumption peaks represent the writes of the encoded video while the first one most likely represents the beginning of reading the file to transcode.

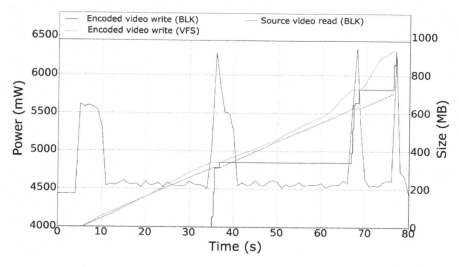

Figure 9.20. *Total power consumption and volume of I/Os during the transcoding process using* PRESET *set to* ultrafast *in a HDD. For a color version of this figure, see www.iste.co.uk/boukhobza/flash.zip*

Figure 9.21 shows the result of the same experiment with an SSD. We can observe a similar behavior from the application (VFS-level writes) and the system (block-level read and writes). On the other hand, the behavior of the power is not the same. In fact, the idle power of SSDs is smaller as well as the power during I/O operations. We can also note the duration of the peaks which is smaller in the SSD case. Another interesting observation that can be made is that the overall duration of the transcoding is the same whether operating on SSD or HDD. This means that for this application, the system is able to correctly pipeline between I/O and CPU processing and that the transcoding is finally a CPU bound application. Furthermore, it also means that for the used application, the storage system is not a bottleneck despite the fact that its optimization results in reducing the power consumption of a few watts.

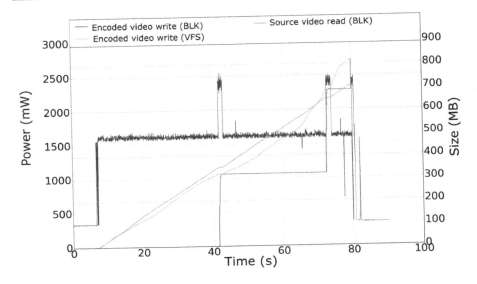

Figure 9.21. *Power consumption and volume of I/Os during the transcoding process using* PRESET *set to* ultrafast *in an SSD. For a color version of this figure, see www.iste.co.uk/boukhobza/flash.zip*

9.4.2.3. *Results with the* PRESET *set to* slow

When the *PRESET slow* is chosen, the first observation that can be made is about the time taken by the transcoding process which is almost 10 times more (see Figure 9.22 and 9.23). As a matter of fact, this value of *PRESET* enables for higher compression to be achieved at the expense of more intensive processing. Concerning the size of the generated file, we note that, on the right of the y-axis for both figures, it is of the same size.

In addition to the difference in the transcoding execution time, we can observe the difference in the activity of the disks. This activity is driven by the flush operations of the kernel. Moreover, it can be noted that the kernel performs much more flushes when the value of the *PRESET* is low. This is explained by the longer duration of the transcoding process. With the *PRESET* set to *ultrafast*, the transcoding lasts about 80 seconds whereas with *slow* it takes 650 seconds. Considering that flushing is carried out every 25 seconds approximately, this yields 26 flushes (650 divided by 25), this is exactly the number of peaks counted during this experiment (not counting the last one which is initiated by *ffmpeg*).

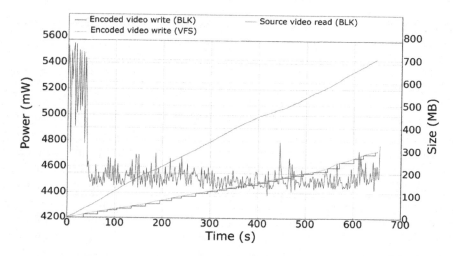

Figure 9.22. *Total power consumption and volume of I/Os during the transcoding process using* PRESET *set to* slow *in a HDD. For a color version of this figure, see www.iste.co.uk/boukhobza/flash.zip*

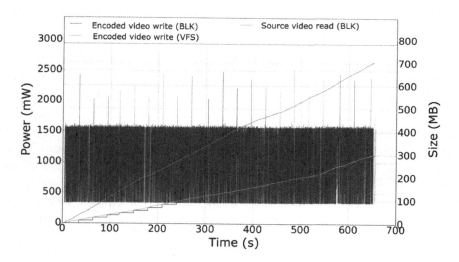

Figure 9.23. *Total power consumption and volume of I/Os during the transcoding process using* PRESET *set to* slow *in an SSD. For a color version of this figure, see www.iste.co.uk/boukhobza/flash.zip*

In this second configuration (*PRESET* set to *slow*), the system is still able of pipelining (time overlapping) between I/Os and CPU operations, such that executing the application takes the same time on both storage devices despite the difference in performance existing between them.

9.5. Conclusions

In this section, we have studied and compared SSDs and HDDs based on three metrics: performance in terms of throughput, power and energy consumption. This has been achieved with micro-benchmarks and actual I/O workloads. We started this chapter with a study on the impact of I/O operations on the energy consumption of computer systems. We have explained that accessing a file on a disk was not limited only to adding the energy consumption of the disk containing the file. As it turns out, not only the consumption of the storage device is added to the rest of the system, but also the processing of all I/O requests by the kernel is included thereto. For this purpose, the latter solicits the CPU and the memory. In addition, during I/O request processing by the storage device, the CPU and the memory consume energy. This consumption turns out to be predominant in the processing of I/O requests and a way to minimize it consists in choosing a fast storage device.

The second part of this chapter has been dedicated to the study of the behavior of HDD and SSD performance and energy consumption. We have noted that both devices behaved in a very different way. Hard disk drives have always been known for achieving a better performance with sequential I/O accesses compared with random ones. This is due to their geometry. The type of operation (read or write) has very little impact on their behavior. On the other hand, SSDs present a very poor homogeneous performance model and which cannot be applied to all flash devices. Generally, there is a difference according to the chosen operation. In the majority of cases, reads perform better than writes and within writes the sequential mode yields the best results. However, this behavior strongly depends on the FTL and its mechanisms. We recall that FTLs are subject to intellectual property and are therefore opaque (unlike FFSs) and that there is no way to grasp their exact functioning. These FTLs are highly variable from one SSD to another. Furthermore, the parallelism used within SSDs is another parameter for differentiation.

The third part shows the results of consumption and power measurements with sensors. When indistinctly considering performance or energy consumption, the SSD behaves much better than the HDD with a performance and a power used 3X to 4X better. It should be reminded that the price per bit is also much higher for SSDs.

PART 4

Emerging Non-volatile Memories

The future belongs to the one who has the longest memory
Friedrich Nietzsche

As a function, memory is as important as arithmetic
Jacques Le Goff

10

Emerging Non-volatile Memories

The objective of this chapter is to present emerging non-volatile memory technologies and to define the principal research avenues of each of them. We will consider four types only: PCM, MRAM, FeRAM and ReRAM, those that appear as the most advanced. After a short introduction to the integration of these emerging memories, we will address them each separately in a section:

1) The first section will be dedicated to the integration of these NVMs.

2) The second section concerns phase-change memory, PCM.

3) The third section describes magneto-resistive memory, MRAM.

4) The fourth section concerns ferroelectric memory, FeRAM.

5) The fifth section describes resistive memories, ReRAM.

10.1. Introduction

Several Non-Volatile Memories (NVM) have emerged in recent years. Among these memories, the most well-known is NAND flash memory whose utilization has exploded in an exponential way. As explained in the introduction, this memory has made it possible to partially close the performance gap that exists between main memory and secondary memory.

Several emerging NVM have been studied in recent years: *Phase-Change Memory* (PCM or also PRAM), *Resistive Memory* (ReRAM), Magneto-Resistive Memory (MRAM), *Ferroelectric Memory* (FeRAM), etc. All these memories have interesting properties which allowed them to be considered in the memory hierarchy pyramid. However, all of them have not reached the same stage of maturity, as it will be described in this chapter.

The characteristics of these NVMs according to the state-of-the-art studies are described in Table 10.1.

10.2. NVM integration

From an architectural and system point of view, a new NVM can be inserted into the memory hierarchy in one of the three conventional positions; namely at the cache processor level (SRAM), at the main memory level (DRAM) or at the secondary memory level (hard disk drive). The integration of these memories can be achieved horizontally or vertically. Recall that horizontal integration means that the NVM is integrated at the same level as existing memory and that during different operations (read or write), the system has the choice regarding the type of memory to use. When vertically integrated, the NVM is positioned between two existing memory levels, it thus shifts one of the two levels and acts as a buffer memory.

10.2.1. *Integration as a storage system*

From the storage system perspective, an NVM can be horizontally or vertically integrated. Similarly to flash memory, some mechanisms for the management of the constraints specific to this NVM must be implemented and abstracted to the highest level layers, in order to enable a simple integration with the conventional management software stack of the storage system. NVMs can also be vertically integrated, as flash memories interfaced by PCIe are. A final possibility is the integration of these memories inside hybrid devices, such as hard disk drives with embedded flash memory chips.

	SRAM	DRAM	HDD	NAND flash	STT-RAM	ReRAM	PCM	FeRAM
Cell size (F^2)	120–200	60–100	N/A	4–6	6–50	4–10	4–12	6–40
Write endurance	10^{16}	$>10^{15}$	$>10^{15}$ (mechanical parts)	10^4–10^5	10^{12}–10^{15}	10^8–10^{11}	10^8–10^9	10^{14}–10^{15}
Read latency	$\sim$0.2–2 ns	$\sim$10 ns	3–5 ms	15–35 μs	2–35 ns	$\sim$10 ns	20–60 ns	20–80 ns
Write latency	$\sim$0.2–2 ns	$\sim$10 ns	3–5 ms	200–500μs	3–50 ns	$\sim$50 ns	20–150 ns	50–75 ns
Leakage power	high	medium	(mechanical parts)	low	low	low	low	low
Dynamic energy (Read/Write)	low	medium	(mechanical parts)	low	low/high	low/high	medium/high	low/high
Maturity	mature	mature	mature	mature	prototypes	prototypes	prototypes	industrialized

Table 10.1. Table summarizing the characteristics of NVMs compared to conventional memories,

[MIT 15, VET 15, XIA 15, WAN 14b, BAE 13, MEE 14, SUR 14]

10.2.2. *Integration as a main memory*

Since numerous NVMs have very interesting performance characteristics (see Table 10.1), they can be integrated to higher levels (than flash memory) of the memory hierarchy, such as main memory. NVMs have performance characteristics rather specific and different from what is usually known for volatile memories such as SRAM and DRAM. It turns out that performances are asymmetric: reads are faster than writes. Endurance may vary in a significant way from one NVM to another and may be further impacted by write operations. Finally, most NVMs make use of electric power only when they are accessed, which is a very interesting property that results in energy saving that can be very significant. All of these characteristics should be taken into account in order to integrate these NVMs as main memory or in conjunction with DRAM to form a hybrid main memory.

At the main memory level, a NVM can also be vertically or horizontally integrated with the DRAM. In the first case, the DRAM can be regarded as a cache for the NVM to reduce access latencies or to mitigate write impact on the endurance of the NVM. The NVM can also be integrated horizontally on the same bus as the DRAM. The management of the placement of the data in any one of the memories can be achieved at three levels: (1) at the hardware level, (2) at the operating system level or (3) at the applicative level. In the first case, memory heterogeneity is hidden from the operating system and from applications, and in the second case it is occulted from applications.

10.2.3. *Integration in CPU caches*

From the point of view of the processor cache, the first cache level is accessed with a very significant frequency. As a result, we can only incorporate therein an NVM with very low latency and very significant endurance that should be comparable to what SRAM offers. This is very difficult, in the current state of the development of NVMs. On the other hand and with respect to the last cache level (L3 or L2 depending on processors) which absorbs a smaller load, some NVMs can be utilized.

Regarding their integration in a processor cache, NVMs must be able to withstand large workloads (compared with integration as DRAM) of write operations. Wear leveling mechanisms must be then employed in order to maintain all the memory usable as long as possible. Furthermore, another important criterion here is compatibility between fabrication processes.

10.3. PCM or phase-change memory

As can be seen in Table 10.1, phase-change memory (the two following abbreviations are in use: PCM or PRAM) has a very small cell physical size (good density), good performance during random access, moderate write throughput, good retention time and a very good endurance compared with NAND flash memory. Unlike flash memory, PCM can be written to (or programmed) without prior erase operation.

10.3.1. *Basic concepts*

PCM uses a chalcogenide-based alloy to implement a memory cell. PCM generally utilizes a thin layer of chalcogenide such as $Ge_2Sb_2Te_5$ (GST) [XUE 11] and two electrodes around the chalcogenide glass in addition to a heat source (see Figure 10.1). PCM is a resistive memory employing different resistance values in order to represent information.

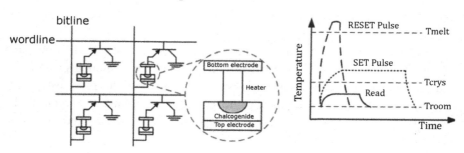

Figure 10.1. *PCM memory cell structure [XUE 11, XIA 15]*

Chalcogenide material quickly changes phase to switch from an amorphous state to a crystalline state. This process is initiated electrically via a heat source. The phase depends on the heat produced by an electrical pulse. A short pulse with a significant voltage results in an amorphous state (RESET, bit set to 0); here, the GST is heated below the melting temperature (T_{melt} in Figure 10.1). Conversely, a long pulse with low voltage yields the crystalline state (SET, bit set to 1). Here, the GST is heated beyond the crystallization temperature (T_{crys} in Figure 10.1), but below the melting temperature. The performance of writes is therefore limited by the longest operation which is SET. Once the phase is established, reads can be performed without disturbing this phase; the ratio between the resistance of the material between the phases SET and RESET is between 10^2 and 10^4. Due to this difference between the resistances of the two states, it is possible to store several bits per cell by making use of intermediate resistances (as is the case for MLC-type flash memories).

Endurance is one of the major problems of PCM memories. As a matter of fact, a cell can only support a maximum number of write operations (usually valued at 10^8). This is mainly because of the heat stress applied in a repetitive manner on the phase-change material. Indeed, thermal expansion and contraction degrades the electrode-storage contact which leads to a decrease of the reliability of programming/writing currents in the cell. The endurance of PCM is better than that of flash (as a reminder, the latter is about 10^5), but much worse than that of DRAM (about 10^{15}). This property is crucial when deciding about the integration of such a technology into the memory hierarchy.

10.3.2. *PCM integration*

The characteristics of write operations decrease the possibilities of using the PCM in the memory hierarchy. It turns out that write latencies as well as endurance do not make it possible to integrate PCM at the processor cache level in view of the very high data traffic that these latter have to endure. In the state-of-the-art, we find horizontally as well as vertically integrated PCMs, as a primary or secondary memory.

10.3.3. *PCM as a main memory*

There are three possibilities for integrating PCM as main memory: (1) as a replacement for DRAM, (2) horizontally on the same level as the DRAM or (3) vertically below DRAM (as a cache). These three possibilities of integration are summarized in Figure 10.2.

In the first and last cases, the existence of PCM is usually abstracted to the operating system while in the second case, it may not be. This means that in some proposals of state-of-the-art architecture, the controller decides of the placement of data in one or the other memory, while in other cases this heterogeneity is exposed to the operating system or to the application that can choose where to place the data. In case the DRAM is used as a cache for the PCM, its objective is to absorb write operations in order to decrease the latency caused by utilizing the PCM and to decrease write cycles sustained by the PCM. Indeed, some experimentation results [QUR 09] have shown that with a DRAM buffer size of about 3% the size of the PCM, it could be possible to obtain a performance and an endurance three times higher than in the case without DRAM. Naturally, it depends on the application workloads and their access pattern.

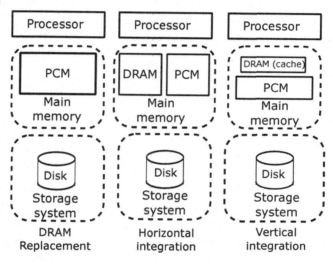

Figure 10.2. *Three architectures for integrating PCMs*

10.3.4. *PCM as a storage system*

Compared with flash memory, PCM has the advantage of allowing random byte access and the possibility to modify data in-place because there are no erase operations to be performed before rewriting such as with flash memory.

PCM prototypes have already emerged such as Onyx [AKE 11] which is a storage system interfaced with PCIe and that has demonstrated better performance than state-of-the-art SSDs for a certain number of access patterns (irregular I/O workloads and dominated by read operations). It should be noted that using the PCIe interface does not enable the optimal utilization of the performance characteristics of this memory because it limits the throughput (PCIe in its first version can provide a theoretical throughput of 4 GB/sec with 16 lines).

Several studies have focused on the vertical integration of PCMs in storage systems. For example, in [SUN 10], PCM is used inside an SSD itself with flash memory in order to absorb the updates to the logs, similar to a cache. The objective is (1) to relieve the flash memory of bursts of updates which decrease the performance as well as the lifetime of flash memory and (2) to improve performance because PCM enables in-place data updates and finally (3) to save energy because PCM is more efficient than DRAM. Other studies also focus on the same principle by integrating PCMs within SSDs in order to replace or supplement the DRAM [LIU 11]. Another objective consists in avoiding data loss in the event of power failure relying on the non-volatility of PCM.

From the operating system point of view, the storage system software stack has been designed for hard disk drives in mind with latencies of the order of microseconds. This is no longer the case with NVMs and efforts must be made in order to adapt the software stack to these new coming NVMs. Bearing this point in mind, several questions arise: how can the new memory hierarchy be abstracted, how can different storage systems be interfaced, what protocols and access mechanisms can be used, etc. As an example, in [PAR 10b], the authors propose two possible integrations for PCM: (1) horizontally with DRAM through a placement achieved depending of the workload access pattern, which would allow large energy savings compared with an implementation using only DRAM. Pages placement is carried out depending on the nature of the memory segments (code, data, etc.) of the executed processes; (2) PCM is employed as additional storage. In this case, randomly accessed data and in small quantities are stored in PCM. In this study, the mechanisms have been implemented in a Linux system with promising results.

10.3.5. *Open questions*

According to [XIA 15], several optimizations have been proposed in state-of-the-art work in order to make this technology viable for different systems.

The main points addressed for optimization are: (1) optimizing write latency, (2) optimizing endurance and finally (3) saving energy:

– Optimizing the write latency: these latencies are due to the physical properties of the phase-change material. The proposed optimizations include: (1) the reduction of the number of written bits, writing only the bits that have actually been modified [ZHO 09b, YAN 07]; (2) the increase of parallelism in writes using the properties of the SET and RESET operations [YUE 13a, DU 13]; (3) the optimization of read operations by giving them priority over write operations that take longer [QUR 10].

– Optimizing the endurance: endurance is a major obstacle in the use of PCM as a replacement for DRAM. There are three endurance optimization classes for PCM. The first solution class consists of acting on electrical properties by optimizing (reducing) the current of the RESET operation thus positively impacting lifetime [JIA 12]. The second solution class consists of minimizing the number of write operations, to this end there are several possibilities: adding a DRAM buffer, avoid rewriting the unmodified bits, avoid writing useless data, compressing the data, etc. Finally, the third optimization class consists of wear leveling. This obviously does not make it possible to decrease the number of writes, but to level wear and thus to maintain the useful space at its maximal value for longer [ZHO 09b, YUN 12, QUR 09].

– Saving energy: again, one of the most relevant arguments in favor for employing NVMs in lieu of the DRAM is their energy efficiency. As a matter of fact, static energy is negligible. Unfortunately, with respect to PCM, the latency in writes is too significant compared with that of DRAM. Several techniques have been proposed in order to address this problem: (1) reducing the number of writes as previously mentioned, (2) reducing the energy consumption of write operations. Here, we can consider the techniques previously mentioned to optimize write latencies. It is also possible to make use of the asymmetry between SET and RESET operations [XU 09, MIR 12, CHE 12, YUE 13b]. It turns out that the RESET operation consumes much more energy than the SET operation.

10.4. MRAM or magneto-resistive memory

One of the most mature and promising NVM technologies is magneto-resistive memory (MRAM). Indeed, MRAM is very dense, exhibits fast write operations, generating low leakage current and having high endurance (see Table 10.1).

10.4.1. *Basic concepts*

MRAM is based on the magnetic properties of a specific material. Unlike DRAM technology, information storage is no longer relative to electric charges but to magnetic orientation. This orientation can be controlled and detected by means of electric currents [KUL 13]. It is not necessary to have an additional external magnetic field to modify the state of the cell.

Magnetic Tunnel Junction (or MTJ) implements the data storage function. It consists of two ferromagnetic layers separated by an insulation layer of oxide (tunnel). The magnetization direction of one of the ferromagnetic layers is fixed and is called reference layer, whereas the other is free and is called free layer (see Figure 10.3).

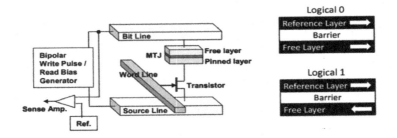

Figure 10.3. *Architecture and operation on MRAM [XUE 11, KUL 13]*

MTJ has low resistance when the two layers have the same magnetization direction. This state is called the parallel state and represents a logical "0" (see Figure 10.3). When the magnetization directions are opposed to one another, this is referred to as an antiparallel state in which the resistance is greater than that of the parallel state. This state represents the logical state "1".

STT-RAM (*Spin Torque Transfer RAM*) or spin transfer RAM is one of the most studied MRAMs. The magnetization direction of the free layer is modified according to the voltage applied to the source line (SL) and bit line (BL) (see Figure 10.3). When a significant positive/negative voltage is applied between SL and BL, a logic "0"/"1" is written. To ensure the change of magnetization direction, the injected current must be maintained for a certain period of time (typically 2ns and 12ns), and its amplitude must be equal to a certain threshold value (usually between 100 μa and 1000 μa).

To perform a read operation, a small voltage is applied between the SL and the BL. This allows current to flow in the MTJ. The value of this current is a function of the resistance of the MTJ. This current is compared with the reference current to be interpreted as a "0" or "1" value read from the MRAM.

A promising MRAM technology is the SOTRAM (*Spin Orbit Torque RAM*) [OBO 15, BIS 14a]. In this technology, the junction utilizes perpendicular magnetization and an additional terminal (or a line) is added such that to separate the path of the read operation from that of the write operation. This implementation allows a better optimization of both paths (reads and writes).

MRAM suffers from two main reliability problems: (1) MTJ cells are prone to thermal instability that can cause loss of data. The probability of failure depends on the implementation of the junction, the temperature, the vulnerability time (time between the write and the last read) and the number of memory cells; (2) The write current is significant, which tends to degrade the junction. This problem affects the STT-RAM cells in which the write current is high. In addition, since the read operation requires current flowing (in the same path), this induces an additional disruption.

10.4.2. *MRAM integration*

As it can be seen in Table 10.1, the endurance of the MRAM is comparable with that of SRAM and that of DRAM. It is thus also a candidate for integration at the same levels of the memory hierarchy that these technologies. However, as noted in [MIT 15], the best real result concerning endurance does not exceed $4 * 10^{12}$. Nonetheless, we will see in the following sections that several efforts have made it possible to increase the endurance of these memories.

From the performance point of view, even if the read performance of STT-RAM is comparable with that of SRAM and may even exceed that of DRAM, the performance and energy properties of writes are not very good.

Most of the state-of-the-art studies integrate MRAM as main memory or cache memory, horizontally or vertically.

10.4.2.1. MRAM as cache

Several state-of-the-art studies focus on using MRAM in different cache levels. Some studies [SEN 14] propose replacing the L1 and L2-level caches with STT-RAM and claim a very significant gain in energy consumption even if performance does not improve for the workloads being tested. Other studies utilize instead a horizontal integration [SUN 09, WU 09, JAD 11, LI 11] by means of which the system places and migrates the data according to the characteristics of workloads with the objective of leveling the problems of significant latencies of write operations.

It is easier to picture the integration of STT-RAM at the last cache level, because performance constraints are weaker at this level (relatively to L1 and L2 levels). An interesting study from [WU 09], regarding the analysis of the integration of NVM, investigates several cache configurations, divided into two groups: (1) *inter-cache Level Hybrid Cache Architectures* (LHCA) and (2) *Region-based Hybrid Cache Architecture* (RHCA). In the first group, each cache level contains a single technology exclusively, the integration is therefore vertical while in the second group, each cache level can contain several regions of different technologies. The study shows that LHCA can enable 7% improvement in IPC (Instructions Per Cycle) while RHCA enables 12%.

10.4.2.2. MRAM as a main memory

Most studies of the integration of MRAM have focused on the cache levels. However, there is a number of existing initiatives on the use of MRAM as a main memory.

In the study conducted by [KUL 13], the authors show that a systematic replacement of DRAM by STT-RAM does not yield very good results, mainly because of the latency of writes. Nevertheless, by performing optimizations specific to MRAM, such as performing partial writes, the circumvention of the *row buffer* (a buffer located within a DRAM memory bank and containing the accessed page), the use of STT-RAM has enabled a significant decrease in energy consumption of about 60%, while maintaining an equivalent performance level.

Another approach by [YAN 13] consists of achieving horizontal integration by implementing a mechanism that makes it possible to direct intensive transactions into DRAM and the rest of the operations to STT-RAM. This is achieved in order to reduce energy consumption without too much impacting on performance.

Although several studies have shown the gain that STT-RAM could contribute according to the mode of integration chosen, in [WAN 14a], the authors particularly highlight the fact that the interfaces used by MRAM and DRAM are not the same. They then propose a few techniques providing an MRAM design compatible with the LPDDR3 format (*Low Power Double Data Rate*).

10.4.3. *Open questions*

In order to be able to include MRAM within a cache level or even as a replacement of DRAM as main memory, several optimizations must be carried out. They are summarized here:

– relaxing the non-volatility of STT-RAM [SMU 11, JOG 12, SUN 11, GUO 10]: the idea is to use the correlation between the retention period and the latency in writes. The shorter the retention period, the smaller the write operation latency will be (and thus energy consumption);

– minimizing the number of write operations [JOO 10, BIS 14b, ZHO 09a, GOS 13, YAZ 14, AHN 12, RAS 10, JUN 13a, PAR 12, SUN 09]: here, as with PCM, the aim is to reduce write latency, for example by writing only modified bits, by encoding the data in order to decrease updates;

– using wear leveling mechanisms;

– managing the asymmetry of writes of bits "0"/"1": the time taken to perform a write achieving the transition from "1" to "0" is more significant than to make the transition from 0 to 1. For instance, resetting all of the cells to "0" before performing a write operation to a word can reduce latency [KWO 14].

10.5. FeRAM or ferroelectric memory

FeRAM (for *Ferroelectric RAM* or FRAM) is an NVM that achieved the stage of industrialization in recent years. As a result, it can be found in different types of micro-controllers [FUJ 13], for example in some systems embedded in cars (wireless boards). FeRAM shows interesting properties that allow it to earn the right to be considered, very early, as an NVM for the future. This is due to a very reduced access latency (of the order of that of the DRAM), reduced energy consumption and a non-erasing write mode (unlike the flash memory), and finally due to the possible direct integration in CMOS technology [KAT 05].

10.5.1. *Basic concepts*

The basic cell of FeRAM memory is similar to that of DRAM, in the sense that it consists of a transistor and a capacitor, except that in this case the capacitor is a ferroelectric one (FCAP). This is thus a cell known as 1T1C (1 transistor, 1 capacitor) [KRY 09, KAN 04]. The ferroelectric material is formed of $Pb(Zr_x,Ti_{1-x})O_3$, also known as lead zirconate titanate or PZT, between two electrodes (see Figure 10.4). FCAP is characterized by two remnant reversible polarization states. The FCAP preserves its polarity without the need for power thanks to the hysteresis property of the ferroelectric material compared with a dielectric material in DRAM. This explains its non-volatility. FeRAM does not need any refresh operation and as a result consumes less power than DRAM.

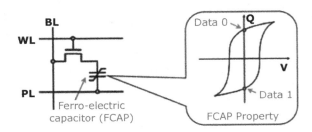

Figure 10.4. *Architecture and operation of a FeRAM cell [JUN 10]*

The write operation is performed by forcing a pulse on the plate line (PL) of the FCAP for a "0" bit and on the bit line (BL) for a bit set to "1". For the FeRAM, there is no need for asymmetric voltage for write operations. The used voltage is the same as V_{cc} and can be very reduced.

FeRAM seemed a very promising NVM as a replacement for DRAM and/or as secondary storage. However, even after several years of study on the technology, FeRAM cells still appear to pose problems when it comes to scaling up. It turns out that their size remains very large (and thus their storage density very low) compared with DRAM for an integration to the gigabit scale. Consequently, the integration of the FeRAM can only be achieved in addition to another memory, or otherwise in small volume in a system.

10.5.2. FeRAM integration

Concerning the performance of FeRAM, several figures have been given in state-of-the-art work. Nevertheless, it seems that most of the studies agree on the fact that the FeRAM, unlike other NVMs, does not present asymmetrical performances between reads and writes. Furthermore, performance seems worse than that of DRAM for better energy consumption. FeRAM offers an endurance between 10^{12} and 10^{15}.

The integration of FeRAMs can be discussed from the perspective of different metrics:

– at the endurance level, FeRAM can replace DRAM or flash-based storage systems;

– at the performance level, FeRAM can complement DRAM (without replacing it in the short-term) or replace flash-based storage memories;

– finally, at the density level, FeRAM cannot replace DRAM nor flash memory because of the size of its cells. As the density of FeRAM has remained low, its cost per bit has remained high in comparison with flash memory.

With regard to its density, FeRAM is not ready to replace current technologies. On the other hand, several studies have been carried out in order to understand the performance and energy benefits if FeRAM were to be integrated between the main memory and the secondary storage.

10.5.2.1. FeRAM as a main memory

Few studies have been conducted assuming FeRAM as a substitute for the DRAM. This is due to the difference of performance between the two technologies.

As an example, in the study from [SUR 14], the authors test several NVMs with different configurations (horizontal and vertical integrations). FeRAM has shown a performance drop whether integrated as a replacement for DRAM or as a complement to it. On the other hand, in the latter case, the system memory proved more efficient energetically speaking, even if performance was worse.

Another interesting study [BAE 13] utilizes a FeRAM prototype and test several possible architectures based on DRAM, flash and other NVMs. The NVM (FeRAM in our case) is placed at different levels (main memory and storage). The authors show that a direct integration without modification gives bad results. A number of optimizations are therefore added, for example the in-place update of data and metadata because it is no longer necessary to have multiple copies between primary and secondary memory since NVMs are non-volatile. The authors also propose a unified memory/storage system per object. These optimizations have enabled considerable improvements in performance. In this study, the authors have also noted the performance issue of FeRAM as well as the problem of density.

10.5.2.2. *FeRAM as a storage system*

Except from the study previously presented [BAE 13] and that of Chameleon [YOO 08] in which FeRAM has been vertically integrated, most state-of-the-art studies make use of FeRAM for horizontal integration in order to optimize the access to the metadata of the file system [JUN 10, DOH 07] or FTL [DOH 07].

In Chameleon, the idea is to store in FeRAM the metadata of the FTL (mapping table) in the FeRAM, and use FeRAM buffers in order to speed up accesses to the underlying flash. In [DOH 07], the authors recommend to store the metadata of the file system rather than storing them in flash memory and load them in RAM during the system initialization. The same idea can be found in the FRASH system [JUN 10]: the concept is to use FeRAM as persistant memory as a means to maintain the metadata of the file system without having to load/unload them from main memory.

10.6. ReRAM or resistive memory

Resistive memories (ReRAM) are memories that store their logical state in the form of a resistive state [SAN 13]. Thanks to their characteristics, resistive memories are supposed to integrate several application domains such as personal computers, medical applications, Internet of Things and automotive applications [MEE 14].

As a matter of fact, resistive memories are compatible with conventional semiconductor fabrication processes of semiconductors. ReRAM has a lower energy consumption than PCMs and greater density than MRAMs (see Table 10.1). ReRAM has also showed good stability with a 10nm technology [AKI 10]. From the endurance perspective, ReRAM has better performance than flash memory (between 1 and 5 orders of magnitude), but remains, however, below the expected endurance for integration as main memory or processor cache. The research effort achieved by several academic and industrial laboratories suggests that many of the issues related to ReRAM could be solved in the near future.

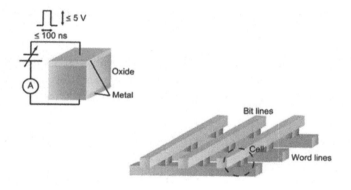

Figure 10.5. *Architecture of a ReRAM cell*[1]

10.6.1. *Basic concepts*

A ReRAM cell, a type of *memristor*, is a passive electronic component with two terminals having a variable electrical resistance. This resistance value changes depending on the magnitude and polarity of the voltage that is applied thereto, as well as the duration of the application [WIL 08]. On a

1 http://www.dobreprogramy.pl/Crossbar-pokazal-terabajtowy-czip-pamieci-ReRAM-niebawem-zapomnimy-o-flashu,News,45456.html.

memristor, the resistance value does not change when the power is turned off, which provides it with the non-volatility property.

The concept of memristor dates from 1971, and was invented by Leon Chua [CHU 71]. The memristor was referenced as the fourth fundamental building block of electronics (at the same level as of the resistor, the capacitor and the inductor). In 2008, a demonstrator has been developed by a team at HP labs [STR 08]: *"Memristor is a contraction of memory resistor, because that is exactly its function: the ability to indefinitely store resistance values means that a memristor can be used as a non-volatile memory"* [WIL 08].

In a memristor, a low-resistance value is considered to be a logical "1" whereas a high-resistance value is considered to be a logic "0" [CLE 14].

A ReRAM cell consists of two metal-insulator-metal (MIM) terminals composed of an insulating or resistive layer (I) sandwiched between two conductive electrodes (M) (see Figure 10.6).

The basic principle of such a cell is the formation and disruption of a conduction filament through the resistive/insulating part (low resistance state) during the application of a sufficiently high voltage (see Figure 10.6).

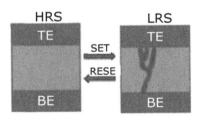

Figure 10.6. *ReRAM cell operation [CLE 14] (HRS for High Resistive State and LRS for Low Resistive State)*

Memristors are one of the most important examples of bipolar device. The switching mechanism is based on the creation of oxygen vacancy that forms an electrical path between the two electrodes. Titanium oxide, or TiO_x, is one of the materials often used but many other materials have been utilized in state-of-the-art work such as ZrO_x [LEE 06], NiO [GIB 64], etc.

10.6.2. *ReRAM integration*

Thanks to the interesting properties of ReRAM (see Table 10.1), they have been considered in several studies as replacement or complementary memories in the memory hierarchy. ReRAM has been studied for vertical and horizontal integrations.

10.6.2.1. *ReRAM as cache*

Even if the endurance of ReRAM memories can appear as an obstacle to its integration in the higher levels of the memory hierarchy, this possibility has been investigated in several state-of-the-art studies.

In [DON 13], the authors have studied the integration of different NVMs and give, in particular, a case study concerning ReRAM. Various performance models have been proposed to allow design space exploration for the integration of ReRAM and other NVMs. These models have been implemented by a simulator called NVSim [DON 12]. The authors of [DON 12] have studied and compared the performance difference with the vertical integration of ReRAM in cache (conventionally SRAM). Numerous scenarios have been tested: placing ReRAM memory in each cache level and then only in the two upper levels (L2 and L3), and finally only in the highest level (L3). For each case, the authors have provided the energy gain and the performance degradation compared with an SRAM memory (the performance of ReRAM are comparable to that of SRAM for reads, but significantly slower for writes). It is thus a matter of compromising between the architecture chosen on the basis of the energy/performance ratio required for the application. In the previous study, the authors have resorted to wear leveling techniques to mitigate the problems of endurance [ZHO 09b, QUR 09, SCH 10b, WAN 13]. The endurance of ReRAM has been considered between 10^{10} and 10^{12} [SHE 11, KIM 11, ESH 11].

ReRAM memory has also been introduced in the memory hierarchy for more specific application domains. In [KOM 13], it has been integrated as a L0 cache (loop buffer [UH 99]) and/or at the L1 cache level for embedded memories targeting ultra-low power wireless multimedia applications with workloads mostly containing read operations. ReRAM memory has also showed good performance when it is integrated into the FPGA in lieu of SRAM to address problems such as leakage currents [CLE 14].

10.6.2.2. ReRAM as a storage system

The endurance of ReRAM does not allow it to occupy a place at the top of the memory hierarchy; as a result, ReRAM memory has been considered as a future competitor or complement to flash memory.

In [TAN 14], the authors have studied a hybrid storage system consisting of a small ReRAM volume and a large flash memory volume. The integration of ReRAM was horizontal with the objective to store parity metadata and error correction code information, ReRAM being non-volatile, enduring and supporting byte-wise updates. The proposed system, called Unified Solid State Storage (USSS), aims to improve the Acceptable Raw Bit Error Rate (ABER).

In [JUN 13b], the authors have highlighted the major problems in designing large-scale crossbar ReRAM such as the disturbance in writes and the read sensing margin [JUN 13b]. The authors have argued that, without taking this issue in consideration, the sensing margin would be unacceptable for large scale ReRAM. The authors have proposed micro-architectural solutions based on a new access policy in order to deal with this problem. They also have proposed a macro-architectural design and have evaluated the performance of 8 GB ReRAM compared with two flash memory configurations (integration as a replacement for flash memory). The proposed architecture has yielded up to 8 times more bandwidth and between 66% and 88% less latency.

Other studies have horizontally integrated ReRAM in an SSD such that to absorb certain specific I/O patterns and thus relieve the flash memory from updates of intensive and random fragmented data [SUN 14] [FUJ 12].

10.6.3. Open questions

Several open research issues related to this memory are discussed in [XUE 11] and [YAN 13]. Some are summarized as follows:

– Endurance is related to the number of cycles that the device can endure before the performance degradation reaches an intolerable level. The endurance values described in the literature are far from what enables the metal-oxide device. The key lies in the choice of switching material; numerous materials have been studied, such as Pt/TaOx/Ta, TiOx, HfOx [LEE 10], WOx [CHI 10]. Some have shown nearly 10 billion endurance switching cycles, which is still far from the endurance sought-after for the universal memory (about 10^{15} to 10^{16}), that is to say the memory technology that could replace all memories of the hierarchy (cache, main and secondary memory).

– Even if the energy needed to switch the state of a ReRAM cell is very small compared with other NVMs such as flash memory or PCM, it is very important to further decrease its value in order to increase memory density. In general, the energy needed for switching (writing) is greater than that for reading. In order to reduce this energy consumption, there are mainly two solutions: (1) lowering the junction size or (2) optimizing the switching material [MIA 11].

– Electroforming is the application of a high voltage or current that must be performed only once in order to enable the switching of the memory cell. This process is responsible for the reduction of electronic conductivity [XUE 11, YAN 09]. The device operates, after that, with large variance properties depending on the material used. It is very important to limit this variation [YAN 09].

– It is possible to have sneak currents, i.e. current flowing through an unexpected path inside the crossbar structure, which can lead to the inadvertently writing or reading of data [JUN 13b]. This is due to the absence of transistor or any other access device in the crossbar structure of the ReRAM memory. This property is necessary for large-scale production [CHE 11b, CHE 11a]. As indicated in [JUN 13b], state-of-the-art solutions can introduce other problems such as the disruption of write operations and the decrease of the sensing margin for reads.

In order to explore the design space of ReRAM memory, many tools have been developed such as NVSim [DON 12]. Furthermore, several models have been developed with the aim to study and optimize numerous parameters such as surface, latency, energy [XU 11], reliability [XU 14] and to study the design of multi-level cells [XU 13].

10.7. Conclusion

In this chapter, we have discussed several promising NVMs that could soon be introduced into the memory hierarchy of computer systems. These NVMs, such as MRAM, FeRAM, ReRAM or PCM, each have their own characteristics, strengths and weaknesses, which makes them able to be inserted at different levels of the memory hierarchy: cache, main memory or storage.

The NVMs being studied are different from memories currently in use. It turns out that they show performance and structure characteristics that are asymmetric. They are addressable by bytes like DRAM and SRAM, but are non-volatile such as flash memory. With respect to performance, they are asymmetric: read operations are more efficient than writes, and even with regard to writes, the fact of writing a "1" or a "0" is achieved with different performances. All of these elements lead the scientific community into investigating the management of these memories in order to integrate them in a profitable manner within current systems.

Conclusion

This book is intended to be an introduction to flash memory, to its integration in current computer systems and also to its behavior. Relying on a primarily experimental methodology, the main focus is on performance and energy consumption.

For this purpose, we have divided this book into four parts. The first part is introductory. The first chapter aims to describe the motivations that have led to writing this book. As a matter of fact, there are very few books dedicated to flash storage technologies and their integration within computer systems. This topic is very relevant at the time when data management is becoming a major societal and economic challenge. The next chapter briefly outlines the basic concepts of flash memory as well as its integration. It introduces the key elements allowing for the comprehension of the other more advanced chapters. Since this book is mainly based on experimental measurements, we have dedicated the third chapter to performance evaluation of storage systems. This chapter describes the different methods and tools for the evaluation of storage systems with a focus on flash memory. This chapter is the basis for the development of the experimental methodologies of the second and third parts of this book.

The second part of the book addresses flash storage systems with dedicated file system (FFS), mainly employed in the area of embedded systems. In the first chapter, we have presented FFSs, their structure and their integration, considering the main three FFSs as case studies: JFFS2, YAFFS and UBIFS. We have highlighted their specificities compared with conventional file systems. Next, a methodology for the analysis of the performance and power consumption of this type of system has been developed. This methodology takes into account the whole management software stack of I/O requests, from the application and to the access to hardware in flash memory. It is based on a multi-level approach, each level being evaluated and analyzed according to lower levels. This analysis has been made possible thanks to the development of tools that enable the profiling of execution traces of I/O requests. Applied to actual embedded boards, this methodology is a means to identify the key elements impacting the performance and the energy consumption of this type of storage system. This has been the subject of the last chapter of this part, in which a set of results is used to help highlight the impact of the different layers of the kernel on the performance and consumption of these systems: VFS (with its cache policies), FFS, MTD and flash memory itself.

The third part is dedicated to flash memory-based storage systems like SSD or other devices for which the integration is performed such that not to change the I/O stack of the kernel. In fact, a particular layer has been developed and inserted within the controller of the storage device, the FTL. It makes it possible to manage the specificities of flash memory and as a result the flash device is simply seen as a block device by the kernel. A chapter has therefore been dedicated to this FTL layer and to the services that it implements. An evaluation of the state-of-the-art mechanisms of address mapping, wear leveling and garbage collection has been conducted. This has contributed to illustrate the fact that a significant difference in performance may exist between the various SSDs on the market. It turns out that the design space of an FTL is very wide and this creates very heterogeneous performance models of SSDs. Subsequently, we have described in a chapter the I/O management stack of the Linux kernel, followed by a methodology for measuring performance and energy consumption. In the next chapter, this methodology has enabled us to understand the impact of the different storage devices on performance and energy consumption of computer systems.

The last part mainly presents non-volatile memory technologies that, most likely, will invade our daily lives in a very near future. These memories are very promising from the performance and energy consumption perspective. They have particular characteristics that drive the scientific community to work on their integration. We have made an attempt to provide an overview of some of these technologies that most likely seem to being industrialized.

Despite a very high cost at its launch, flash technology has been able to find its place in the memory hierarchy mainly due to the boom of smartphones sales. It invades increasingly more our daily lives: USB sticks, SD cards for cameras or micro SD for telephones, TVs, cameras and various connected objects, etc. Its use is increasing even more due to its performance and energy properties in data centers, despite a higher price than traditional hard disk drives. As we have seen through the various chapters of this book, flash memory presents very good performance, easily reaching a 3X difference factor when compared with a hard disk. To accompany the boom of the volume of digital data, we have a growing need for more storage volume so as to absorb all this information. In this context, energy and space economy along with performance optimization are all the more arguments in favor of the development of this type of memory and other technologies such as those introduced in Chapter 10.

In order to take advantage of flash memory, consistent work has begun to be carried out to make the I/O management software stack in the kernels more efficient. Indeed, due to its very large design space, flash memory can offer very good performance, relying on parallelism, for example. The I/O stack must be relieved for this performance not to be hampered. This work is all the more important to integrate new non-volatile memories.

A quick look at any vendor site allows us to calculate the difference of price that may exist between two SSDs, a priori of the same capacity. This is quite different from what was known of conventional storage systems. Although there has always been a difference in price between hard disk drives for workstations and those for PCs, the difference was mainly due to the rotation speed (7200, 10k or 15k), to the cache size and then to the storage density. For a SSD, more factors are taken into consideration: the cell type (SLC, MLC, TLC), the complexity of the FTL being used and the performance of the controller, the cache memory, the parallelism implemented, etc. All these elements drive the cautious buyer into consulting the technical details prior to purchase. Obviously, the cost will depend on the desired performance and as such, one may note the various interfaces that drives based on flash memory can use. Furthermore, depending on the complexity of the architecture, the flash device can offer a too significant throughput to be interfaced by SATA for example. It turns out that SATA III supports speeds of about 600 MB/s only. In this case, the PCIe interface can be used and other interfaces such as NVMe are under study.

Besides the cost, flash memories present a number of guaranteed advantages in terms of performance and energy consumption. On the other hand, bad dimensioning can lead to a degradation of performance which can be very significant. Overloading an SSD or having too many write solicitations without timeout can result in a much lower performance due to the garbage collector being initiated. Thus, a poor choice of the type of SSD with respect to the writing volume that it can absorb may lead to its early failure and cause data loss. In addition, it can be difficult to optimize the performance of a given application for SSD because the performance model is not homogeneous as it was the case for hard drives, even though general tendencies may be expressed (see Chapter 9).

Flash memories have changed our perspective of storage, and other technological revolutions are awaiting in a very near future. As described in Chapter 10, there are several candidate NVMs upon which a close watch should be kept. Several manufacturers such as Intel/Micron or even Samsung have already announced the production of NVM-based storage devices. Major changes must be made to existing systems to take advantage of these memories, in order to be able to quickly access them, to minimize the energy footprint of the storage and to secure the access. Extraordinary scientific challenges are being presented to us and it is paramount to address them because in a world where all our data are digitized, storage systems emerge as the guardians of our collective memory.

Bibliography

[ACH 99] ACHIWA K., YAMAMOTO A., YAMAGATA H., Memory system using a flash memory and method of controlling the memory system, US Patent 5,930,193, 27 July 1999.

[AHN 12] AHN J., CHOI K., "Lower-bits cache for low power STT-RAM caches", *2012 IEEE International Symposium on Circuits and Systems*, pp. 480–483, May 2012.

[AKE 11] AKEL A., CAULFIELD A.M., MOLLOV T.I. *et al.*, "Onyx: a protoype phase change memory storage array", *Proceedings of the 3rd USENIX Conference on Hot Topics in Storage and File Systems*, HotStorage'11, USENIX Association, Berkeley, CA, p. 2, 2011.

[AKI 10] AKINAGA H., SHIMA H., "Resistive random access memory (ReRAM) based on metal oxides", *Proceedings of the IEEE*, vol. 98, no. 12, pp. 2237–2251, December 2010.

[AND 02] ANDERSON D., CHASE J., Fstress: a flexible network file service benchmark, Report , Technical Report CS-2002-01, Duke University, Department of Computer Science, 2002.

[ARA 04] ARANYA A., WRIGHT C.P., ZADOK E., "Tracefs: a file system to trace them all", *FAST*, pp. 129–145, 2004.

[ASS 95] ASSAR M., NEMAZIE S., ESTAKHRI P., Flash memory mass storage architecture incorporation wear leveling technique, US Patent 5,479,638, 26 December 1995.

[AXB 14] AXBOE J., Fio documentation, available at: https://github.com/axboe/fio/blob/master/HOWTO, 2014.

[AXI 99] AXIS COMMUNICATIONS, JFFS Home Page, available at: http://developer.axis.com/ old/software/jffs/, 1999.

[BAE 13] BAEK S., CHOI J., LEE D. *et al.*, "Energy-efficient and high-performance software architecture for storage class memory", *ACM Transactions on Embedded Computing Systems*, vol. 12, no. 3, pp. 81:1–81:22, ACM, April 2013.

[BAN 95] BAN A., Flash file system, US Patent 5404485, April 1995.

[BAN 99] BAN A., Flash file system optimized for page-mode flash technologies, US Patent 5,937,425, 10 August 1999.

[BAR 95] BARRETT P.L., QUINN S.D., LIPE R.A., System for updating data stored on a flash-erasable, programmable, read-only memory (FEPROM) based upon predetermined bit value of indicating pointers, US Patent 5392427, February 1995.

[BEL 05] BELLARD F., "QEMU, a fast and portable dynamic translator", *USENIX Annual Technical Conference, FREENIX Track*, pp. 41–46, 2005.

[BEN 14] BENMOUSSA Y., SENN E., BOUKHOBZA J., "Open-PEOPLE, a collaborative platform for remote and accurate measurement and evaluation of embedded systems power consumption", *22nd IEEE International Symposium on Modelling, Analysis and Simulation of Computer and Telecommunication Systems*, 2014.

[BEZ 03] BEZ R., CAMERLENGHI E., MODELLI A. *et al.*, "Introduction to flash memory", *Proceedings of the IEEE*, pp. 489–502, 2003.

[BIS 14a] BISHNOI R., EBRAHIMI M., OBORIL F. *et al.*, "Architectural aspects in design and analysis of SOT-based memories", *19th Asia and South Pacific Design Automation Conference (ASP-DAC)*, pp. 700–707, January 2014.

[BIS 14b] BISHNOI R., OBORIL F., EBRAHIMI M. *et al.*, "Avoiding unnecessary write operations in STT-MRAM for low power implementation", *Fifteenth International Symposium on Quality Electronic Design*, pp. 548–553, March 2014.

[BIT 05] BITYUTSKIY A., JFFS3 Design Issues, available at: http://www.linux-mtd.infradead.org/doc/JFFS3design.pdf, 2005.

[BJØ 10] BJØRLING M., BONNET P., BOUGANIM L. *et al.*, "uFLIP: understanding the energy consumption of flash devices", *IEEE Data Engineering Bulletin*, vol. 33, no. 4, pp. 48–54, 2010.

[BOU 09] BOUGANIM L., JÓNSSON B., BONNET P., "uFLIP: understanding flash IO patterns", arXiv Preprint arXiv:0909.1780, 2009.

[BOU 11a] BOUGANIM L., BONNET P., "Flash device support for database management", *Proceedings of the 5th Biennial Conference on Innovative Data System*, Asilomar, CA, pp. 1–8, 2011.

[BOU 11b] BOUKHOBZA J., OLIVIER P., RUBINI S., "A cache management strategy to replace wear leveling techniques for embedded flash memory", *Performance Evaluation of Computer & Telecommunication Systems (SPECTS), 2011 International Symposium on*, IEEE, pp. 1–8, 2011.

[BOU 13a] BOUKHOBZA J., OLIVIER P., RUBINI S., "CACH-FTL: a cache-aware configurable hybrid flash translation layer", *2013 21st Euromicro International Conference on Parallel, Distributed, and Network-Based Processing*, pp. 94–101, February 2013.

[BOU 13b] BOUKHOBZA J., "Flashing in the cloud: shedding some light on NAND flash memory storage systems", *Data Intensive Storage Services for Cloud Environments*, pp. 241–266, IGI Global, April 2013.

[BOU 15] BOUKHOBZA J., OLIVIER P., RUBINI S. *et al.*, "MaCACH: an adaptive cache-aware hybrid FTL mapping scheme using feedback control for efficient page-mapped space management", *Journal of Systems Architecture – Embedded Systems Design*, vol. 61, no. 3–4, pp. 157–171, 2015.

[BOV 05] BOVET D.P., CESATI M., *Understanding the Linux Kernel*, O'Reilly Media, 2005.

[BRA 96] BRAY T., The Bonnie file system benchmark, available at: http://www.textuality.com/bonnie/, 1996.

[BRO 12] BROWN N., "An F2FS teardown", *Linux Weekly News*, available at: https://lwn.net/Articles/518988/, 2012.

[BRU 07] BRUNELLE A.D., Blktrace user guide, available at: http://www.cse.unsw.edu.au/aaronc/iosched/doc/blktrace.html, 2007.

[BRY 02] BRYANT R., FORESTER R., HAWKES J., "Filesystem performance and scalability in linux 2.4. 17", *Proceedings of the Freenix Track: 2002 USENIX Annual Technical Conference*, USENIX Association, pp. 259–274, 2002.

[BUC 08] BUCY J.S., SCHINDLER J., SCHLOSSER S.W. *et al.*, The disksim simulation environment version 4.0 reference manual (cmu-pdl-08-101), Report, Parallel Data Laboratory, 2008.

[CAR 10] CARTER J., RAJAMANI K., "Designing energy-efficient servers and data centers", *Computer*, vol. 43, no. 7, pp. 76–78, July 2010.

[CHA 02] CHANG L., KUO T., "An adaptive striping architecture for flash memory storage systems of embedded systems", *Proceedings of the 8th IEEE Conference on Real-Time and Embedded Technology and Applications Symposium, 2002*, pp. 187–196, 2002.

[CHA 05] CHANG R., QAWAMI B., SABET-SHARGHI F., Method and apparatus for managing an erase count block, EP Patent 20, 030,752,155, 3 August 2005.

[CHA 07] CHANG L.-P., "On efficient wear leveling for large-scale flash-memory storage systems", *Proceedings of the 2007 ACM Symposium on Applied Computing*, New York, NY, ACM, pp. 1126–1130, 2007.

[CHE 07] CHEN J.J., MIELKE N.R., CALVINHU C., "Flash memory reliability", in BREWER J., GILL M., (eds.), *Nonvolatile Memory Technologies with Emphasis on Flash*, John Wiley & Sons, 2007.

[CHE 11a] CHEN Y.C., LI H., CHEN Y. *et al.*, "3D-ICML: a 3D bipolar ReRAM design with interleaved complementary memory layers", *2011 Design, Automation Test in Europe*, pp. 1–4, March 2011.

[CHE 11b] CHEN Y.C., LI H., ZHANG W. *et al.*, "3D-HIM: a 3D high-density interleaved memory for bipolar RRAM design", *2011 IEEE/ACM International Symposium on Nanoscale Architectures*, pp. 59–64, June 2011.

[CHE 12] CHEN J., CHIANG R.C., HUANG H.H. *et al.*, "Energy-aware writes to non-volatile main memory", *SIGOPS Operating Systems Review*, vol. 45, no. 3, pp. 48–52, ACM, January 2012.

[CHI 99a] CHIANG M.L., CHANG R.C., "Cleaning policies in mobile computers using flash memory", *Journal of Systems and Software*, vol. 48, no. 3, pp. 213–231, November 1999.

[CHI 99b] CHIANG M.-L., LEE P.C.H., CHANG R.-C., "Using data clustering to improve cleaning performance for flash memory", *Software: Practice and Experience*, vol. 29, no. 3, pp. 267–290, 1999.

[CHI 08] CHIANG M.L., CHENG C.L., WU C.H., "A new FTL-based flash memory management scheme with fast cleaning mechanism", *International Conference on Embedded Software and Systems, 2008. ICESS'08*, pp. 205–214, July 2008.

[CHI 10] CHIEN W.C., CHEN Y.C., LAI E.K. *et al.*, "Unipolar switching behaviors of RTO WOx RRAM", *IEEE Electron Device Letters*, vol. 31, no. 2, pp. 126–128, February 2010.

[CHI 11] CHIAO M., CHANG D., "ROSE: a novel flash translation layer for NAND flash memory based on hybrid address translation", *IEEE Transactions on Computers*, vol. 60, no. 6, pp. 753–766, 2011.

[CHO 09] CHO H., SHIN D., EOM Y.I., "KAST: K-associative sector translation for NAND flash memory in real-time systems", *2009 Design, Automation Test in Europe Conference Exhibition*, pp. 507–512, April 2009.

[CHU 71] CHUA L., "Memristor-the missing circuit element", *IEEE Transactions on Circuit Theory*, vol. 18, no. 5, pp. 507–519, September 1971.

[CLE 14] CLERMIDY F., JOVANOVIC N., ONKARAIAH S. *et al.*, "Resistive memories: which applications?", *2014 Design, Automation Test in Europe Conference Exhibition (DATE)*, pp. 1–6, March 2014.

[COK 09] COKER R., The bonnie++ benchmark, available at: http://www.coker.com.au/bonnie+, 2009.

[COO 09] COOPERSTEIN D.J., *Writing Linux Device Drivers: Lab Solutions a Guide with Exercises*, CreateSpace, Paramount, 2009.

[COR 05] CORBET J., RUBINI A., KROAH-HARTMAN G., *Linux Device Drivers*, 3rd ed., O'Reilly Media, 2005.

[DAV 10] DAVIS J.D., RIVOIRE S., Building energy-efficient systems for sequential workloads, Technical Report MSR-TR-2010-30, Microsoft Research, 2010.

[DAY 13] DAYAN N., SVENDSEN M.K.E., BJÖRLING M. *et al.*, "EagleTree: exploring the design space of SSD-based algorithms", *Proceedings of the VLDB Endowment*, vol. 6, no. 12, pp. 1290–1293, 2013.

[DEB 09] DEBNATH B., SUBRAMANYA S., DU D. *et al.*, "Large block CLOCK (LB-CLOCK): a write caching algorithm for solid state disks", *2009 IEEE International Symposium on Modeling, Analysis Simulation of Computer and Telecommunication Systems*, pp. 1–9, September 2009.

[DOH 07] DOH I.H., CHOI J., LEE D. *et al.*, "Exploiting non-volatile RAM to enhance flash file system performance", *Proceedings of the 7th ACM - IEEE International Conference on Embedded Software*, New York, pp. 164–173, 2007.

[DON 12] DONG X., XU C., XIE Y. *et al.*, "Nvsim: a circuit-level performance, energy, and area model for emerging nonvolatile memory", *IEEE Transactions on Computer-Aided Design of Integrated Circuits and Systems*, vol. 31, no. 7, pp. 994–1007, 2012.

[DON 13] DONG X., JOUPPI N.P., XIE Y., "A circuit-architecture co-optimization framework for evaluating emerging memory hierarchies", *IEEE International Symposium on Performance Analysis of Systems and Software (ISPASS)*, pp. 140–141, April 2013.

[DOU 96] DOUGLIS F., CACERES R., KAASHOEK M.F. *et al.*, "Storage alternatives for mobile computers", *Mobile Computing*, vol. 353, pp. 473–505, 1996.

[DU 13] DU Y., ZHOU M., CHILDERS B.R. *et al.*, "Bit mapping for balanced PCM cell programming", *Proceedings of the 40th Annual International Symposium on Computer Architecture*, New York, pp. 428–439, 2013.

[EIG 05] EIGLER F.C., PRASAD V., COHEN W. *et al.*, Architecture of systemtap: A Linux trace/probe tool, available at: https://sourceware.org/systemtap/archpaper.pdf, 2005.

[ELM 10] ELMAGHRAOUI K., KANDIRAJU G., JANN J. *et al.*, "Modeling and simulating flash based solid-state disks for operating systems", *Proceedings of the First Joint WOSP/SIPEW International Conference on Performance Engineering*, pp. 15–26, 2010.

[ENG 05] ENGEL J., MERTENS R., "LogFS-finally a scalable flash file system", *12th International Linux System Technology Conference*, 2005.

[ESH 11] ESHRAGHIAN K., CHO K.R., KAVEHEI O. *et al.*, "Memristor MOS content addressable memory (MCAM): hybrid architecture for future high performance search engines", *IEEE Transactions on Very Large Scale Integration (VLSI) Systems*, vol. 19, no. 8, pp. 1407–1417, August 2011.

[FAR 10] FARMER D., Study projects nearly 45-fold annual data growth by 2020, available at: http://www.passmarksecurity.com/about/news/press/2010/20100504-01.htm, 2010.

[FAZ 07] FAZIO A., BAUER M., "Multilevel cell digital memories", in BREWER J., GILL M. (eds), *Nonvolatile Memory Technologies with Emphasis on Flash*, John Wiley & Sons, 2007.

[FIC 11] FICHEUX P., *Linux Embarqué*, Blanche, Eyrolles, 2011.

[FOR 07] FORNI G., ONG C., RICE C. *et al.*, "Flash memory applications", in BREWER J., GILL M. (eds), *Nonvolatile Memory Technologies with Emphasis on Flash*, John Wiley & Sons, 2007.

[FUJ 12] FUJII H., MIYAJI K., JOHGUCHI K. *et al.*, "x11 performance increase, x6.9 endurance enhancement, 93% energy reduction of 3D TSV-integrated hybrid ReRAM/MLC NAND SSDs by data fragmentation suppression", *Symposium on VLSI Circuits (VLSIC)*, pp. 134–135, June 2012.

[FUJ 13] FUJISAKI Y., "Review of emerging new solid-state non-volatile memories", *Japanese Journal of Applied Physics*, vol. 52, no. 4R, p. 040001, 2013.

[GIB 64] GIBBONS J., BEADLE W., "Switching properties of thin Nio films", *Solid-State Electronics*, vol. 7, no. 11, pp. 785–790, 1964.

[GLE 06a] GLEDITSCH A.G., GJERMSHUS P.K., Linux cross reference, available at: https://sourceforge.net/projects/lxr/, 2006.

[GLE 06b] GLEIXNER T., HAVERKAMP F., BITYUTSKIY A., UBI-unsorted block images, Report, 2006.

[GOS 13] GOSWAMI N., CAO B., LI T., "Power-performance co-optimization of throughput core architecture using resistive memory", *IEEE 19th International Symposium on High Performance Computer Architecture (HPCA2013)*, pp. 342–353, February 2013.

[GRU 09] GRUPP L.M., CAULFIELD A.M., COBURN J. *et al.*, "Characterizing flash memory: anomalies, observations, and applications", *42nd Annual IEEE/ACM International Symposium on Microarchitecture*, pp. 24–33, 2009.

[GUA 13] GUAN Y., WANG G., WANG Y. *et al.*, "BLog: block-level log-block management for NAND flash memorystorage systems", *Proceedings of the 14th ACM SIGPLAN/SIGBED Conference on Languages, Compilers and Tools for Embedded Systems*, New York, pp. 111–120, 2013.

[GUO 10] GUO X., IPEK E., SOYATA T., "Resistive computation: avoiding the power wall with low-leakage, STT-MRAM based computing", *SIGARCH Computer Architecture News*, vol. 38, no. 3, pp. 371–382, June 2010.

[GUP 09] GUPTA A., KIM Y., URGAONKAR B., "DFTL: a flash translation layer employing demand-based selective caching of page-level address mappings", *Proceedings of the 14th International Conference on Architectural Support for Programming Languages and Operating Systems*, New York, pp. 229–240, 2009.

[GUR 09] GURUMURTHI S., "Architecting storage for the cloud computing era", *IEEE Micro*, vol. 29, no. 6, pp. 68–71, 2009.

[HAV 11] HAVASI F., "An improved B+ tree for flash file systems", *SOFSEM 2011: Theory and Practice of Computer Science*, pp. 297–307, 2011.

[HEG 08] HEGER D., JACOBS J., RAO S. *et al.*, The flexible file system benchmark webpage, available at: https://sourceforge.net/projects/ffsb/, 2008.

[HOE 09] HOELZLE U., BARROSO L.A., *The Datacenter as a Computer: an Introduction to the Design of Warehouse-Scale Machines*, Morgan and Claypool Publishers, 2009.

[HOM 09] HOMMA T., "Evaluation of Flash File Systems for Large NAND Flash Memory", *CELF Embedded Linux Conference*, San Francisco, 6th–8th April 2009.

[HOR 14] HOROWITZ M., "1.1 Computing's energy problem (and what we can do about it)", *IEEE International Solid-State Circuits Conference Digest of Technical Papers (ISSCC)*, pp. 10–14, February 2014.

[HOW 88] HOWARD J.H., KAZAR M.L., MENEES S.G. *et al.*, "Scale and performance in a distributed file system", *ACM Transactions on Computer Systems (TOCS)*, vol. 6, no. 1, pp. 51–81, 1988.

[HP 99] HP LABS, Cello 99 Traces – SNIA, available at: http://iotta.snia.org/traces/21, 1999.

[HSI 08] HSIEH J.W., TSAI Y.L., KUO T.W. *et al.*, "Configurable flash-memory management: performance versus overheads", *IEEE Transactions on Computers*, vol. 57, no. 11, pp. 1571–1583, November 2008.

[HU 09] HU X., ELEFTHERIOU E., HAAS R. *et al.*, "Write amplification analysis in flash-based solid state drives", *Proceedings of SYSTOR 2009: The Israeli Experimental Systems Conference*, p. 10, 2009.

[HU 10] Hu J., Jiang H., Tian L. *et al.*, "PUD-LRU: an erase-efficient write buffer management algorithm for flash memory SSD", *IEEE International Symposium on Modeling, Analysis and Simulation of Computer and Telecommunication Systems*, pp. 69–78, August 2010.

[HU 11] Hu Y., Jiang H., Feng D. *et al.*, "Performance impact and interplay of SSD parallelism through advanced commands, allocation strategy and data granularity", *Proceedings of the International Conference on Supercomputing*, pp. 96–107, 2011.

[HUN 08] Hunter A., A brief introduction to the design of UBIFS, Report, available at: http://www.linux-mtd.infradead.org/doc/ubifs_whitepaper.pdf, March 2008.

[HWA 12] Hwang J.-Y., "F2FS: A new file system designed for flash storage in mobile", *Embedded Linux Conference Europe*, Barcelona, Spain, 2012.

[INC 13] Inc M., "TN-FD-20: C400 SSD SMART Attributes Introduction", SMART Attributes for the C400 SSD, available at: https://www.micron.com/media/documents/products/technical-note/solid-state-storage/tnfd20_c400_ssd_smart_attributes.pdf, 2013.

[INT 08] Intel Corp. and Micron Inc., Intel and Micron develop the world's fastest NAND flash memory with 5X faster performance, available at: http://www.intc.com/releasedetail.cfm?ReleaseID=311576, 2008.

[INT 09] Intel Corp., Intel X25-M Solid State Drive Datasheet, available at: http://download.intel.com/design/flash/nand/mainstream/mainstream-sata-ssd-datasheet.pdf, 2009.

[JAD 11] Jadidi A., Arjomand M., Sarbazi-Azad H., "High-endurance and performance-efficient design of hybrid cache architectures through adaptive line replacement", *International Symposium on Low Power Electronics and Design (ISLPED)*, pp. 79–84, August 2011.

[JAI 08] Jain R., *The Art of Computer Systems Performance Analysis*, John Wiley & Sons, 2008.

[JIA 11] Jiang S., Zhang L., Yuan X. *et al.*, "S-FTL: an efficient address translation for flash memory by exploiting spatial locality", *2011 IEEE 27th Symposium on Mass Storage Systems and Technologies (MSST)*, pp. 1–12, May 2011.

[JIA 12] Jiang L., Zhang Y., Yang J., "ER: elastic RESET for low power and long endurance MLC based phase change memory", *Proceedings of the 2012 ACM/IEEE International Symposium on Low Power Electronics and Design*, New York, pp. 39–44, 2012.

[JO 06] Jo H., Kang J.-U., Park S.-Y. *et al.*, "FAB: flash-aware buffer management policy for portable media players", *IEEE Transactions on Consumer Electronics*, vol. 52, no. 2, pp. 485–493, May 2006.

[JOG 12] Jog A., Mishra A.K., Xu C. *et al.*, "Cache revive: architecting volatile STT-RAM caches for enhanced performance in CMPs", *Proceedings of the 49th Annual Design Automation Conference*, New York, pp. 243–252, 2012.

[JOH 13] Johnson N., Jetstress Field Guide, Report, Microsoft, available at: https://gallery.technet.microsoft.com/office/Jetstress-2013-Field-Guide-2438bc12, 2013.

[JOO 10] JOO Y., NIU D., DONG X. *et al.*, "Energy- and endurance-aware design of phase change memory caches", *2010 Design, Automation Test in Europe Conference Exhibition (DATE 2010)*, pp. 136–141, March 2010.

[JOU 05] JOUKOV N., WONG T., ZADOK E., "Accurate and Efficient Replaying of File System Traces", *FAST*, vol. 5, pp. 25–25, 2005

[JUN 08] JUNG D., KIM J., KIM J. *et al.*, "ScaleFFS: a scalable log-structured flash file system for mobile multimedia systems", *ACM Transactions on Multimedia Computing, Communications, and Applications (TOMCCAP)*, vol. 5, no. 1, p. 9, 2008.

[JUN 10] JUNG J., WON Y., KIM E. *et al.*, "FRASH: exploiting storage class memory in hybrid file system for hierarchical storage", *Transactions on Storage*, vol. 6, no. 1, pp. 3:1–3:25, ACM, April 2010.

[JUN 12] JUNG M., WILSON E.H., DONOFRIO D. *et al.*, "NANDFlashSim: intrinsic latency variation aware NAND flash memory system modeling and simulation at microarchitecture level", *IEEE 28th Symposium on Mass Storage Systems and Technologies (MSST)*, IEEE, pp. 1–12, 2012.

[JUN 13a] JUNG J., NAKATA Y., YOSHIMOTO M. *et al.*, "Energy-efficient spin-transfer torque RAM cache exploiting additional all-zero-data flags", *14th International Symposium on Quality Electronic Design (ISQED)*, pp. 216–222, March 2013.

[JUN 13b] JUNG M., SHALF J., KANDEMIR M., "Design of a large-scale storage-class RRAM system", *Proceedings of the 27th International ACM Conference on International Conference on Supercomputing*, New York, pp. 103–114, 2013.

[KAN 04] KANAYA H., TOMIOKA K., MATSUSHITA T. *et al.*, "A 0.602 μm^2 nestled 'Chain' cell structure formed by one mask etching process for 64 Mbit FeRAM", *Symposium on VLSI Technology, 2004. Digest of Technical Papers*, pp. 150–151, June 2004.

[KAN 09] KANG Y., MILLER E.L., "Adding aggressive error correction to a high-performance compressing flash file system", *Proceedings of the Seventh ACM International Conference on Embedded Software*, New York, pp. 305–314, 2009.

[KAT 97] KATCHER J., Postmark: a new file system benchmark, Technical Report TR3022, Network Appliance, 1997, available at: www. netapp. com/tech_library/3022. html, 1997.

[KAT 05] KATO Y., YAMADA T., SHIMADA Y., "0.18-μm nondestructive readout FeRAM using charge compensation technique", *IEEE Transactions on Electron Devices*, vol. 52, no. 12, pp. 2616–2621, December 2005.

[KAW 95] KAWAGUCHI A., NISHIOKA S., MOTODA H., "A flash-memory based file system", *Proceedings of the USENIX 1995 Technical Conference Proceedings*, Berkeley, CA, pp. 13–13, 1995.

[KEN 14] KENISTON J., PANCHAMUKHI P.S., HIRAMATSU M., Linux Kernel probes documentation, available at: https://www.kernel.org/doc/Documentation/kprobes.txt, 2014.

[KIM 02] KIM J., KIM J.M., NOH S.H. *et al.*, "A space-efficient flash translation layer for CompactFlash systems", *IEEE Transactions on Consumer Electronics*, vol. 48, no. 2, pp. 366–375, May 2002.

[KIM 08] KIM H., AHN S., "BPLRU: a buffer management scheme for improving random writes in flash storage", *FAST*, vol. 8, pp. 1–14, 2008.

[KIM 09a] KIM H., WON Y., KANG S., "Embedded NAND flash file system for mobile multimedia devices", *IEEE Transactions on Consumer Electronics*, vol. 55, no. 2, pp. 545–552, 2009.

[KIM 09b] KIM Y., TAURAS B., GUPTA A. *et al.*, "Flashsim: a simulator for nand flash-based solid-state drives", *First International Conference on Advances in System Simulation*, pp. 125–131, 2009.

[KIM 11] KIM Y.B., LEE S.R., LEE D. *et al.*, "Bi-layered RRAM with unlimited endurance and extremely uniform switching", *VLSI Technology (VLSIT), 2011 Symposium on*, pp. 52–53, June 2011.

[KIM 12a] KIM H., AGRAWAL N., UNGUREANU C., "Revisiting storage for smartphones", *ACM Transactions on Storage (TOS)*, vol. 8, no. 4, p. 14, 2012.

[KIM 12b] KIM J., SHIM H., PARK S. *et al.*, "FlashLight: a lightweight flash file system for embedded systems", *ACM Transactions on Embedded Computer Systems*, vol. 11S, no. 1,Pages 18:1–18:23, June 2012.

[KIM 12c] KIM J., KIM J., "Androbench: benchmarking the storage performance of Android-based mobile devices", *Frontiers in Computer Education*, vol. 133, pp. 667–674, Springer, 2012.

[KOM 13] KOMALAN M.P., PÉREZ J. I.G., TENLLADO C. *et al.*, "Design exploration of a NVM based hybrid instruction memory organization for embedded platforms", *Design Automation for Embedded Systems*, vol. 17, no. 3, pp. 459–483, 2013.

[KOP 12] KOPYTOV A., "SysBench manual", MySQL AB, available at: http://imysql.com/wp-content/uploads/2014/10/sysbench-manual.pdf, 2012.

[KRY 09] KRYDER M.H., KIM C.S., "After hard drives – what comes next?", *IEEE Transactions on Magnetics*, vol. 45, no. 10, pp. 3406–3413, October 2009.

[KUL 13] KULTURSAY E., KANDEMIR M., SIVASUBRAMANIAM A. *et al.*, "Evaluating STT-RAM as an energy-efficient main memory alternative", *IEEE International Symposium on Performance Analysis of Systems and Software (ISPASS)*, pp. 256–267, April 2013.

[KUO 02] KUOPPALA M., Tiobench-Threaded I/O bench for Linux, available at: https://sourceforge.net/projects/tiobench/, 2002.

[KWO 07] KWON O., KOH K., "Swap-aware garbage collection for NAND flash memory based embedded systems", *7th IEEE International Conference on Computer and Information Technology*, pp. 787–792, October 2007.

[KWO 11] KWON S.J., RANJITKAR A., KO Y.-B. *et al.*, "FTL algorithms for NAND-type flash memories", *Design Automation for Embedded Systems*, vol. 15, no. 3–4, pp. 191–224, 2011.

[KWO 14] KWON K.W., CHODAY S.H., KIM Y. *et al.*, "AWARE (Asymmetric Write Architecture With REdundant Blocks): a high write speed STT-MRAM cache architecture", *IEEE Transactions on Very Large Scale Integration (VLSI) Systems*, vol. 22, no. 4, pp. 712–720, April 2014.

[LAY 10] LAYTON, J.B., "Harping on metadata performance: new benchmarks", *Linux Magazine*, available at: http://www.linux-mag.com/id/7742/, 2010.

[LEE 06] LEE D., SEONG D.J., CHOI H.J. *et al.*, "Excellent uniformity and reproducible resistance switching characteristics of doped binary metal oxides for non-volatile resistance memory applications", *International Electron Devices Meeting*, pp. 1–4, December 2006.

[LEE 07] LEE S.-W., PARK D.-J., CHUNG T.-S. *et al.*, "A log buffer-based flash translation layer using fully-associative sector translation", *ACM Transactions on Embedded Computing Systems*, vol. 6, no. 3, ACM, July 2007.

[LEE 08] LEE S., SHIN D., KIM Y.-J. *et al.*, "LAST: locality-aware sector translation for NAND flash memory-based storage systems", *SIGOPS Operating Systems Review*, vol. 42, no. 6, pp. 36–42, ACM, October 2008.

[LEE 09a] LEE H.S., YUN H.S., LEE D.H., "HFTL: hybrid flash translation layer based on hot data identification for flash memory", *IEEE Transactions on Consumer Electronics*, vol. 55, no. 4, pp. 2005–2011, November 2009.

[LEE 09b] LEE S., MOON B., PARK C., "Advances in flash memory SSD technology for enterprise database applications", *Proceedings of the 2009 ACM SIGMOD International Conference on Management of Data*, ACM, pp. 863–870, 2009.

[LEE 10] LEE H.Y., CHEN Y.S., CHEN P.S. *et al.*, "Evidence and solution of over-RESET problem for HfOX based resistive memory with sub-ns switching speed and high endurance", *IEEE International Electron Devices Meeting (IEDM)*, pp. 19.7.1–19.7.4, December 2010.

[LEE 11] LEE C., LIM S., "Caching and deferred write of metadata for Yaffs2 flash file system", *9th International Conference on Embedded and Ubiquitous Computing (EUC)*, pp. 41–46, 2011.

[LEE 12] LEE K., WON Y., "Smart layers and dumb result: IO characterization of an Android-based smartphone", *Proceedings of the Tenth ACM International Conference on Embedded Software*, pp. 23–32, 2012.

[LEE 14] LEE S., KIM J., "Improving performance and capacity of flash storage devices by exploiting heterogeneity of MLC flash memory", *IEEE Transactions on Computers*, vol. 63, no. 10, pp. 2445–2458, 2014.

[LEN 13] LENSING P.H., CORTES T., BRINKMANN A., "Direct lookup and hash-based metadata placement for local file systems", *Proceedings of the 6th International Systems and Storage Conference*, p. 5, 2013.

[LEV 04] LEVON J., *OProfile Manual*, Victoria University of Manchester, available at: http://oprofile.sourceforge.net/doc/, 2004.

[LI 11] LI J., XUE C.J., XU Y., "STT-RAM based energy-efficiency hybrid cache for CMPs", *2011 IEEE/IFIP 19th International Conference on VLSI and System-on-Chip*, pp. 31–36, October 2011.

[LIM 06] LIM S., PARK K., "An efficient NAND flash file system for flash memory storage", *IEEE Transactions on Computers*, vol. 55, no. 7, pp. 906–912, 2006.

[LIU 10] LIU S., GUAN X., TONG D. *et al.*, "Analysis and comparison of NAND flash specific file systems", *Chinese Journal of Electronics*, vol. 19, no. 3, pp. 403–408, 2010.

[LIU 11] LIU Y., ZHOU C., CHENG X., "Hybrid SSD with PCM", *11th Annual Non-Volatile Memory Technology Symposium (NVMTS)*, pp. 1–5, November 2011.

[LOV 10] LOVE R., *Linux Kernel Development*, 3rd ed., Addison-Wesley Professional, 2010.

[MAN 10] MANNING C., How YAFFS works, available at: http://www.yaffs.net/ documents/how-yaffs-works, 2010.

[MAT 09] MATHUR G., DESNOYERS P., CHUKIU P. *et al.*, "Ultra-low power data storage for sensor networks", *ACM Transactions on Sensor Networks (TOSN)*, vol. 5, no. 4, p. 33, 2009.

[MEE 14] MEENA J.S., SZE S.M., CHAND U. *et al.*, "Overview of emerging nonvolatile memory technologies", *Nanoscale Research Letters*, vol. 9, no. 1, pp. 1–33, 2014.

[MEM 01] MEMIK G., MANGIONE-SMITH W.H., HU W., "Netbench: a benchmarking suite for network processors", *Proceedings of the 2001 IEEE/ACM International Conference on Computer-Aided Design*, pp. 39–42, 2001.

[MES 11] MESNIER M., Intel open storage toolkit, available at: https://sourceforge. net/projects/intel-iscsi/, 2011.

[MIA 11] MIAO F., STRACHAN J.P., YANG J.J. *et al.*, "Anatomy of a nanoscale conduction channel reveals the mechanism of a high-performance memristor", *Advanced Materials*, vol. 23, no. 47, pp. 5633–5640, 2011.

[MIC 04] MICRON INC., NAND flash performance increase using the micron PAGE READ CACHE MODE command, Technical report, available at: https://www.micron. com/media/Documents/Products/Technical%20Note/NAND%20Flash/tn2901.pdf, 2004.

[MIC 05] MICRON INC., Small-Block vs. Large-Block NAND Flash Devices, Report no. TN-29-07, available at: https://www.micron.com/~/media/documents/products/technical-note/nand-flash/tn2907.pdf, 2005.

[MIC 06] MICRON INC., NAND flash performance increase with PROGRAM PAGE CACHE MODE command, Technical report, available at: https://www.micron. com/~/media/documents/products/technical-note/nand-flash/tn2914.pdf, 2006.

[MIC 09a] MICRON INC., NAND flash and mobile LPDDR 168-Ball package-on-package (PoP) MCP combination memory (TI OMAP) datasheet, available at: https://4donline.ihs.com/images/VipMasterIC/IC/MICT/MICTS02342/MICTS02402-1.pdf, 2009.

[MIC 09b] MICROSOFT RESEARCH, SSD Extension for DiskSim simulation environment, available at: https://www.microsoft.com/en-us/download/details.aspx?id=52332, 2009.

[MIC 10] MICRON INC., Micron NAND flash memory x8, x16 NAND flash memory features, available at: https://www.micron.com/~/media/products/ data-sheet/nand-flash/20-series/2gb_nand_m29b.pdf, 2010.

[MIR 12] MIRHOSEINI A., POTKONJAK M., KOUSHANFAR F., "Coding-based energy minimization for phase change memory", *Proceedings of the 49th Annual Design Automation Conference*, New York, pp. 68–76, 2012.

[MIS 13] MISTRAL SOLUTIONS, OMAP35x Evaluation module, available at: https://www.mistralsolutions.com/product-engineering-services/pes-library/?action=downloadproduct&id=1141, 2013.

[MIT 15] MITTAL S., VETTER J.S., LI D., "A survey of architectural approaches for managing embedded DRAM and non-volatile on-chip caches", *IEEE Transactions on Parallel and Distributed Systems*, vol. 26, no. 6, pp. 1524–1537, June 2015.

[MOH 10] MOHAN V., Modeling the physical characteristics of NAND FLASH memory, PhD Thesis, University of Virginia, 2010.

[MTD 05] MTD CONTRIBUTORS, JFFS2 documentation – MTD website, available at: http://www.linux-mtd.infradead.org/doc/jffs2.html, 2005.

[MTD 08] MTD CONTRIBUTORS, NandSim Linux flash simulator, available at: http://www.linux-mtd.infradead.org/faq/nand.html#L_nand_nandsim, 2008.

[MTD 09] MTD CONTRIBUTORS, MTD general documentation – can I mount ext2 over an MTD device – MTD website, available at: http://www.linux-mtd.infradead.org/faq/general.html#L_ext2_mtd, 2009.

[NAT 14] NATIONAL INSTRUMENTS, PXI-4472B dynamic signal acquisition module, available at: http://sine.ni.com/nips/cds/view/p/lang/en/nid/12184, 2014.

[NOR 03] NORCOTT W.D., CAPPS D., Iozone filesystem benchmark, available at: http://www.iozone.org/docs/IOzone_msword_98.pdf, 2003.

[OBO 15] OBORIL F., BISHNOI R., EBRAHIMI M. *et al.*, "Evaluation of hybrid memory technologies using SOT-MRAM for on-chip cache hierarchy", *IEEE Transactions on Computer-Aided Design of Integrated Circuits and Systems*, vol. 34, no. 3, pp. 367–380, March 2015.

[OLI 12] OLIVIER P., BOUKHOBZA J., SENN E., "Micro-benchmarking flash memory file-system wear leveling and garbage collection: a focus on initial state impact", *Proceedings of the 2012 IEEE 15th International Conference on Computational Science and Engineering*, Washington DC, pp. 437–444, 2012.

[OLI 13] OLIVIER P., BOUKHOBZA J., SENN E., "Modeling driver level NAND flash memory I/O performance and power consumption for embedded Linux", *2013 11th International Symposium on Programming and Systems (ISPS)*, pp. 143–152, 2013.

[OLI 14a] OLIVIER P., BOUKHOBZA J., SENN E., "Flashmon V2: monitoring raw NAND flash memory I/O requests on embedded linux", *ACM SIGBED Review*, vol. 11, no. 1, pp. 38–43, 2014.

[OLI 14b] OLIVIER P., BOUKHOBZA J., SENN E., "Revisiting read-ahead efficiency for raw NAND flash storage in embedded Linux", *Proceedings of the 4th Embed With Linux (EWiLi) Workshop*, 2014.

[OLI 14c] OLIVIER P., BOUKHOBZA J., SOULA M. *et al.*, "A tracing toolset for embedded Linux flash file systems", *Proceedings of the 8th International Conference on Performance Evaluation Methodologies and Tools (VALUETOOLS)*, Bratislava, Slovakia, pp. 153–158, 2014.

[OLI 16] OLIVIER P., BOUKHOBZA J., SENN E. *et al.*, "A Methodology for estimating performance and power consumption of embedded flash file systems", *ACM Transactions on Embedded Computing Systems (TECS)*, vol. 15, no. 4, p. 79, ACM, 2016.

[ONF 14] ONFI WORKGROUP, Open NAND flash interface specification – Revision 4.0, Report , available at: http://www.onfi.org/~/media/onfi/specs/onfi_4_0-gold.pdf?la=en, 2014.

[OPD 10] OPDENACKER M., Flash filesystem benchmarks, available at: http://elinux.org/Flash_Filesystem_Benchmarks, 2010.

[OPE 12] OPEN PEOPLE CONTRIBUTORS, Site web Open-PEOPLE website, available at: https://www.open-people.fr, 2012.

[OPE 13] OPENBENCHMARKING.COM CONTRIBUTORS, Aio-stress test profile – OpenBenchmarking.com, available at: https://openbenchmarking.org/test/pts/aio-stress, 2013.

[OUA 14] OUARNOUGHI H., BOUKHOBZA J., SINGHOFF F. *et al.*, "A multi-level I/O tracer for timing and performance storage systems in IaaS cloud", *3rd IEEE International Workshop on Real-time and Distributed Computing in Emerging Applications*, Rome, Italy, 2 December 2014.

[PAR 90] PARK A., BECKER J.C., LIPTON R.J., "IOstone: a synthetic file system benchmark", *ACM SIGARCH Computer Architecture News*, vol. 18, no. 2, pp. 45–52, 1990.

[PAR 06a] PARK S.-Y., JUNG D., KANG J.-U. *et al.*, "CFLRU: a replacement algorithm for flash memory", *Proceedings of the 2006 International Conference on Compilers, Architecture and Synthesis for Embedded Systems*, pp. 234–241, 2006.

[PAR 06b] PARK S., LEE T., CHUNG K., "A flash file system to support fast mounting for NAND flash memory based embedded systems", *Embedded Computer Systems: Architectures, Modeling, and Simulation*, pp. 415–424, 2006.

[PAR 08] PARK Y., LIM S., LEE C. *et al.*, "PFFS: a scalable flash memory file system for the hybrid architecture of phase-change RAM and NAND flash", *Proceedings of the 2008 ACM symposium On Applied Computing*, pp. 1498–1503, 2008.

[PAR 09] PARK J., YOO S., LEE S. *et al.*, "Power Modeling of Solid State Disk for Dynamic Power Management Policy Design in Embedded Systems", in LEE S., NARASIMHAN P., (eds.), *Software Technologies for Embedded and Ubiquitous Systems*, Springer Berlin Heidelberg, January 2009.

[PAR 10a] PARK D., DEBNATH B., DU D., "CFTL: a convertible flash translation layer adaptive to data access patterns", *Proceedings of the ACM SIGMETRICS International Conference on Measurement and Modeling of Computer Systems*, New York, pp. 365–366, 2010.

[PAR 10b] PARK Y., PARK S.K., PARK K.H., "Linux kernel support to exploit phase change memory", *Proceedings of the Ottawa Linux Symposium*, 2010.

[PAR 11] PARK S., KIM Y., URGAONKAR B. *et al.*, "A comprehensive study of energy efficiency and performance of flash-based SSD", *Journal of System Architecture*, vol. 57, no. 4, pp. 354–365, 2011.

[PAR 12] PARK S.P., GUPTA S., MOJUMDER N. *et al.*, "Future cache design using STT MRAMs for improved energy efficiency: devices, circuits and architecture", *49th ACM/EDAC/IEEE Design Automation Conference (DAC)*, pp. 492–497, June 2012.

[PAR 13] PARK S.O., KIM S.J., "ENFFiS: An enhanced NAND flash memory file system for mobile embedded multimedia system", *ACM Transactions on Embeded Computer Systems*, vol. 12, no. 2, pp. 23:1–23:13, February 2013.

[QIN 10] QIN Z., WANG Y., LIU D. *et al.*, "Demand-based block-level address mapping in large-scale NAND flash storage systems", *Proceedings of the Eighth IEEE/ACM/IFIP International Conference on Hardware/Software Codesign and System Synthesis*, New York, pp. 173–182, 2010.

[QIN 11] QIN Z., WANG Y., LIU D. *et al.*, "MNFTL: an efficient flash translation layer for MLC NAND flash memory storage systems", *Proceedings of the 48th Design Automation Conference*, New York, pp. 17–22, 2011.

[QUR 09] QURESHI M.K., SRINIVASAN V., RIVERS J.A., "Scalable high performance main memory system using phase-change memory technology", *Proceedings of the 36th Annual International Symposium on Computer Architecture*, New York, pp. 24–33, 2009.

[QUR 10] QURESHI M.K., FRANCESCHINI M.M., LASTRAS-MONTANO L.A., "Improving read performance of phase change memories via write cancellation and write pausing", *HPCA – 16 2010 The Sixteenth International Symposium on High-Performance Computer Architecture*, pp. 1–11, January 2010.

[RAN 11] RANGANATHAN P., "From microprocessors to nanostores: rethinking data-centric systems", *IEEE Computer*, vol. 44, no. 1, pp. 39–48, 2011.

[RAS 10] RASQUINHA M., CHOUDHARY D., CHATTERJEE S. *et al.*, "An energy efficient cache design using spin torque transfer (STT) RAM", *ACM/IEEE International Symposium on Low-Power Electronics and Design (ISLPED)*, pp. 389–394, August 2010.

[REA 12] REARDON J., CAPKUN S., BASIN D.A., "Data node encrypted file system: efficient secure deletion for flash memory", *USENIX Security Symposium*, pp. 333–348, 2012.

[RIC 14a] RICHTER D., "Fundamentals of non-volatile memories", *Flash Memories*, no. 40, pp. 5–110, January 2014.

[RIC 14b] RICHTER D., "Fundamentals of reliability for flash memories", *Flash Memories*, no. 40, pp. 149–166, January 2014.

[ROB 09] ROBERTS D., KGIL T., MUDGE T., "Using non-volatile memory to save energy in servers", *2009 Design, Automation Test in Europe Conference Exhibition*, pp. 743–748, April 2009.

[ROO 08] ROOHPARVAR F.F., Single level cell programming in a multiple level cell non-volatile memory device, US Patent 7366013, April 2008.

[ROS 92] ROSENBLUM M., OUSTERHOUT J.K., "The design and implementation of a log-structured file system", *ACM Transactions on Computer Systems (TOCS)*, vol. 10, no. 1, pp. 26–52, 1992.

[ROS 08] ROSTEDT S., Ftrace documentation, US Patent, available at: http://www.google.ch/patents/US7366013, 2008.

[RUS 06] RUSSINOVICH M., COGSWELL B., FileMon for Windows v7. 04, available at: https://technet.microsoft.com/en-us/sysinternals/filemon.aspx, 2006.

[SAK 07] SAKUI K., SUH K.-D., "NAND flash memory technology", in BREWER J., GILL M. (eds), *Nonvolatile Memory Technologies with Emphasis on Flash*, John Wiley & Sons, 2007.

[SAM 07] SAMSUNG ELECTRONICS CO., Samsung K9XXG08XXM NAND Flash chip datasheet, available at: http://www.dataman.com/media/datasheet/Samsung/K9WBG08U1M_K9KAG08U0M_K9NCG08U5M_rev10.pdf, 2007.

[SAM 09] SAMSUNG ELECTRONICS CO., Samsung PM800 solid state drive datasheet, available at: http://www.bdtic.com/DATASHEET/SAMSUNG/MMDOE56G5MXP-0VB_MMCRE28G5MXP-0VB_MMCRE64G5MXP-0VB.PDF, 2009.

[SAN 13] SANDHU G.S., "Emerging memories technology landscape", *13th Non-Volatile Memory Technology Symposium (NVMTS)*, pp. 1–5, August 2013.

[SCH 09] SCHIERL A., SCHELLHORN G., HANEBERG D. *et al.*, "Abstract specification of the UBIFS file system for flash memory", *FM 2009: Formal Methods*, pp. 190–206, 2009.

[SCH 10a] SCHALL D., HUDLET V., HÄRDER T., "Enhancing energy efficiency of database applications using SSDs", *Proceedings of the Third C* Conference on Computer Science and Software Engineering*, pp. 1–9, 2010.

[SCH 10b] SCHECHTER S., LOH G.H., STRAUSS K. *et al.*, "Use ECP, Not ECC, for hard failures in resistive memories", *SIGARCH Computer Architecture News*, vol. 38, no. 3, pp. 141–152, June 2010.

[SEN 12] SENN E., CHILLET D., ZENDRA O. *et al.*, "Open-people: open power and energy optimization PLatform and estimator", *15th Euromicro Conference on Digital System Design (DSD)*, pp. 668–675, 2012.

[SEN 14] SENNI S., TORRES L., SASSATELLI G. *et al.*, "Exploration of Magnetic RAM Based Memory Hierarchy for Multicore Architecture", *2014 IEEE Computer Society Annual Symposium on VLSI*, pp. 248–251, July 2014.

[SEO 08] SEO E., PARK S., URGAONKAR B., "Empirical analysis on energy efficiency of flash-based SSDs.", *Workshop on Power Aware Computing and Systems (HotPower '08)*, 2008.

[SHE 11] SHEU S.S., CHANG M.F., LIN K.F. *et al.*, "A 4Mb embedded SLC resistive-RAM macro with 7.2ns read-write random-access time and 160ns MLC-access capability", *2011 IEEE International Solid-State Circuits Conference*, pp. 200–202, February 2011.

[SHI 99] SHINOHARA T., Flash memory card with block memory address arrangement, US Patent 5905993, available at: https://www.google.com/patents/US5905993, May 1999.

[SHI 10] SHIN D., "Power Analysis for flash memory SSD", *Workshop for Operating System Support for Non-Volatile RAM (NVRAMOS)*, Jeju, Korea, 2010.

[SMU 11] SMULLEN C.W., MOHAN V., NIGAM A. *et al.*, "Relaxing non-volatility for fast and energy-efficient STT-RAM caches", *2011 IEEE 17th International Symposium on High Performance Computer Architecture*, pp. 50–61, February 2011.

[SNI 11] SNIA, Storage network industry association IOTTA trace repository, available at: http://iotta.snia.org/, 2011.

[SPC 13] SPC CONTRIBUTORS, SPC Benchmark 1 specifications, Report, available at: http://www.storageperformance.org/specs/SPC-1_SPC-1E_v1.14.pdf, 2013.

[SPE 08] SPEC CONTRIBUTORS, SPECSFS2008 User Guide, Report, available at: https://www.spec.org/sfs2008/docs/usersguide.pdf, 2008.

[ST 04] ST MICROELECTRONICS, Bad block management in NAND flash memories, Application Note no. AN1819, Technical report, available at: http://www.eetasia.com/ARTICLES/2004NOV/A/2004NOV29_MEM_AN06.PDF?SOURCES=DOWNLOAD, 2004.

[STA 91] STALLMAN R.M., PESCH R.H., Using GDB: a guide to the GNU source-level debugger, available at: https://www.eecs.umich.edu/courses/eecs373/readings/Debugger.pdf, 1991.

[STE 85] STEFFEN J.L. *et al.*, "Interactive examination of a C program with Cscope", *Proceedings of the USENIX Winter Conference*, pp. 170–175, 1985.

[STR 08] STRUKOV D.B., SNIDER G.S., STEWART D.R. *et al.*, "The missing memristor found", *Nature*, vol. 453, no. 7191, pp. 80–83, 2008.

[SU 09] SU X., JIN P., XIANG X. *et al.*, "Flash-DBSim: a simulation tool for evaluating flash-based database algorithms", *2nd IEEE International Conference on Computer Science and Information Technology*, pp. 185–189, August 2009.

[SUN 09] SUN G., DONG X., XIE Y. *et al.*, "A novel architecture of the 3D stacked MRAM L2 cache for CMPs", *2009 IEEE 15th International Symposium on High Performance Computer Architecture*, pp. 239–249, February 2009.

[SUN 10] SUN G., JOO Y., CHEN Y. *et al.*, "A hybrid solid-state storage architecture for the performance, energy consumption, and lifetime improvement", *Sixteenth International Symposium on High-Performance Computer Architecture*, pp. 1–12, January 2010.

[SUN 11] SUN Z., BI X., LI H.H. *et al.*, "Multi retention level STT-RAM cache designs with a dynamic refresh scheme", *Proceedings of the 44th Annual IEEE/ACM International Symposium on Microarchitecture*, New York, pp. 329–338, 2011.

[SUN 14] SUN C., MIYAJI K., JOHGUCHI K. *et al.*, "A high performance and energy-efficient cold data eviction algorithm for 3D-TSV hybrid ReRAM/MLC NAND SSD", *IEEE Transactions on Circuits and Systems I: Regular Papers*, vol. 61, no. 2, pp. 382–392, February 2014.

[SUR 14] SURESH A., CICOTTI P., CARRINGTON L., "Evaluation of emerging memory technologies for HPC, data intensive applications", *2014 IEEE International Conference on Cluster Computing (CLUSTER)*, pp. 239–247, September 2014.

[SYU 05] SYU S.-J., CHEN J., "An active space recycling mechanism for flash storage systems in real-time application environment", *11th IEEE International Conference on Embedded and Real-Time Computing Systems and Applications (RTCSA'05)*, pp. 53–59, August 2005.

[TAN 14] TANAKAMARU S., DOI M., TAKEUCHI K., "NAND flash memory/ReRAM hybrid unified solid-state-storage architecture", *IEEE Transactions on Circuits and Systems I: Regular Papers*, vol. 61, no. 4, pp. 1119–1132, April 2014.

[TEH 13] TEHRANI S., "Advancement in charge-trap flash memory technology", *2013 5th IEEE International Memory Workshop*, 2013.

[TOS 09] TOSHIBA INC., Evaluation of UBI and UBIFS, available at: http://elinux.org/images/f/f8/CELFJamboree30-UBIFS_update.pdf, 2009.

[TPC 14] TPC CONTRIBUTORS, Transaction Performance Council Benchmarks Webpage, available at: http://www.tpc.org/information/benchmarks.asp, 2014.

[TRA 08] TRAEGER A., ZADOK E., JOUKOV N. *et al.*, "A nine year study of file system and storage benchmarking", *ACM Transactions on Storage (TOS)*, vol. 4, no. 2, Page5, 2008.

[TRI 08] TRIDGELL A., DBench Benchmark documentation, available at: https://dbench.samba.org/doc/dbench.1.html, 2008.

[TS' 05] TS'O T., The EXT2FS library, available at: http://www.giis.co.in/libext2fs.pdf, 2005.

[UBI 09a] UBI DEVELOPERS, MTD Website – UBI Documentation – NAND Flash sub-pages, available at: http://www.linux-mtd.infradead.org/doc/ubi.html#L_subpage, 2009.

[UBI 09b] UBI DEVELOPERS, UBI FAQ – MTD website, available at: http://www.linux-mtd.infradead.org/faq/ubi.html, 2009.

[UH 99] UH G.-R., WANG Y., WHALLEY D. *et al.*, "Effective exploitation of a zero overhead loop buffer", *Proceedings of the ACM SIGPLAN 1999 Workshop on Languages, Compilers, and Tools for Embedded Systems*, New York, pp. 10–19, 1999.

[UMA 09] UMASS, UMASS (University of Massachusetts) Storage Trace Repository, available at: http://traces.cs.umass.edu/, 2009.

[VAN 12] VANDENBERGH H., Vdbench Users Guide, available at: http://www.oracle.com/technetwork/server-storage/vdbench-1901683.pdf, 2012.

[VAS 10] VASUDEVAN V., ANDERSEN D., KAMINSKY M. *et al.*, "Energy-efficient cluster computing with FAWN: workloads and implications", *Proceedings of the 1st International Conference on Energy-Efficient Computing and Networking*, New York, pp. 195–204, 2010.

[VEN 08] VENKATESWARAN S., *Essential Linux Device Drivers*, Prentice Hall Press, Upper Saddle River, 2008.

[VET 15] VETTER J.S., MITTAL S., "Opportunities for nonvolatile memory systems in extreme-scale high-performance computing", *Computing in Science Engineering*, vol. 17, no. 2, pp. 73–82, March 2015.

[WAN 08] WANG Y., SHU J., XUE W. *et al.*, "VFS interceptor: dynamically tracing file system operations in real environments", *First International Workshop on Storage and I/O Virtualization, Performance, Energy, Evaluation and Dependability*, 2008.

[WAN 13] WANG J., DONG X., XIE Y. *et al.*, "i2WAP: improving non-volatile cache lifetime by reducing inter- and intra-set write variations", *IEEE 19th International Symposium on High Performance Computer Architecture (HPCA2013)*, pp. 234–245, February 2013.

[WAN 14a] WANG J., DONG X., XIE Y., "Building and optimizing MRAM-based commodity memories", *ACM Transactions on Architecture and Code Optimization*, vol. 11, no. 4, pp. 36:1–36:22, December 2014.

[WAN 14b] WANG J., DONG X., XIE Y. *et al.*, "Endurance-aware cache line management for non-volatile caches", *ACM Transactions on Architecture and Code Optimization*, vol. 11, no. 1, pp. 4:1–4:25, February 2014.

[WEI 11] WEI Q., GONG B., PATHAK S. *et al.*, "WAFTL: a workload adaptive flash translation layer with data partition", *IEEE 27th Symposium on Mass Storage Systems and Technologies (MSST)*, pp. 1–12, May 2011.

[WIL 08] WILLIAMS R.S., "How we found the missing memristor", *IEEE Spectrum*, vol. 45, no. 12, pp. 28–35, December 2008.

[WIN 08] WINTER R., Why are data warehouses growing so fast?, available at: http://searchdatamanagement.techtarget.com/news/2240111227/Why-Are-Data-Warehouses- Growing-So-Fast, 2008.

[WOL 89] WOLMAN B., OLSON T.M., "IOBENCH: a system independent IO benchmark", *SIGARCH Computer Architecture News*, vol. 17, no. 5, pp. 55–70, September 1989.

[WOO 01] WOODHOUSE D., "JFFS2: the journalling flash file system version 2", *Ottawa Linux Symposium*, Ottawa, 2001.

[WOO 07] WOOKEY, YAFFS – a NAND flash file system, available at: http://wookware.org/talks/yaffscelf2007.pdf, 2007.

[WU 94] WU M., ZWAENEPOEL W., "eNVy: a non-volatile, main memory storage system", *Proceedings of the Sixth International Conference on Architectural Support for Programming Languages and Operating Systems*, New York, pp. 86–97, 1994.

[WU 09] WU X., LI J., ZHANG L. *et al.*, "Hybrid cache architecture with disparate memory technologies", *Proceedings of the 36th Annual International Symposium on Computer Architecture*, New York, pp. 34–45, 2009.

[WU 10] WU G., ECKART B., HE X., "BPAC: an adaptive write buffer management scheme for flash-based solid state drives", *2010 IEEE 26th Symposium on Mass Storage Systems and Technologies (MSST)*, pp. 1–6, May 2010.

[XIA 15] XIA F., JIANG D.-J., XIONG J. *et al.*, "A survey of phase change memory systems", *Journal of Computer Science and Technology*, vol. 30, no. 1, pp. 121–144, 2015.

[XU 09] XU W., LIU J., ZHANG T., "Data manipulation techniques to reduce phase change memory write energy", *Proceedings of the 2009 ACM/IEEE International Symposium on Low Power Electronics and Design*, New York, pp. 237–242, 2009.

[XU 11] XU C., DONG X., JOUPPI N.P. *et al.*, "Design implications of memristor-based RRAM cross-point structures", *2011 Design, Automation Test in Europe*, pp. 1–6, March 2011.

[XU 13] XU C., NIU D., MURALIMANOHAR N. *et al.*, "Understanding the trade-offs in multi-level cell ReRAM memory design", *Design Automation Conference (DAC), 2013 50th ACM/EDAC/IEEE*, pp. 1–6, May 2013.

[XU 14] XU C., NIU D., ZHENG Y. *et al.*, "Reliability-aware cross-point resistive memory design", *Proceedings of the 24th Edition of the Great Lakes Symposium on VLSI*, New York, pp. 145–150, 2014.

[XUE 11] XUE C.J., SUN G., ZHANG Y. *et al.*, "Emerging non-volatile memories: opportunities and challenges", *Proceedings of the 9th International Conference on Hardware/Software Codesign and System Synthesis (CODES+ISSS)*, pp. 325–334, October 2011.

[YAF 12] YAFFS2 CONTRIBUTORS, Google android – YAFFS2 website, available at: http://www.yaffs.net/google-android, 2012.

[YAN 07] YANG B.D., LEE J.E., KIM J.S. *et al.*, "A low power phase-change random access memory using a data-comparison write scheme", *2007 IEEE International Symposium on Circuits and Systems*, pp. 3014–3017, May 2007.

[YAN 09] YANG J.J., MIAO F., PICKETT M.D. *et al.*, "The mechanism of electroforming of metal oxide memristive switches", *Nanotechnology*, vol. 20, no. 21, p. 215201, 2009.

[YAN 13] YANG J.J., WILLIAMS R.S., "Memristive devices in computing system: promises and challenges", *Journal of Emerging Technologies in Computing Systems*, vol. 9, no. 2, pp. 11:1–11:20, ACM, May 2013.

[YAZ 14] YAZDANSHENAS S., PIRBASTI M.R., FAZELI M. *et al.*, "Coding last level STT-RAM cache for high endurance and low power", *IEEE Computer Architecture Letters*, vol. 13, no. 2, pp. 73–76, July 2014.

[YOO 08] YOON J.H., NAM E.H., SEONG Y.J. *et al.*, "Chameleon: a high performance flash/FRAM hybrid solid state disk architecture", *IEEE Computer Architecture Letters*, vol. 7, no. 1, pp. 17–20, January 2008.

[YOO 11a] YOO B., WON Y., CHOI J. *et al.*, "SSD characterization: from energy consumption's perspective", *Proceedings of the 3rd USENIX Conference on Hot Topics in Storage and File Systems*, Berkeley, CA, 2011.

[YOO 11b] YOO S., PARK C., "Low power mobile storage: SSD case study", *Energy-Aware System Design*, pp. 223–246, 2011.

[YUE 13a] YUE J., ZHU Y., "Accelerating write by exploiting PCM asymmetries", *IEEE 19th International Symposium on High Performance Computer Architecture (HPCA2013)*, pp. 282–293, February 2013.

[YUE 13b] YUE J., ZHU Y., "Exploiting subarrays inside a bank to improve phase change memory performance", *Design, Automation Test in Europe Conference Exhibition (DATE), 2013*, pp. 386–391, March 2013.

[YUN 12] YUN J., LEE S., YOO S., "Bloom filter-based dynamic wear leveling for phase-change RAM", *2012 Design, Automation Test in Europe Conference Exhibition (DATE)*, pp. 1513–1518, March 2012.

[ZHA 10] ZHANG Q., CHENG L., BOUTABA R., "Cloud computing: state-of-the-art and research challenges", *Journal of Internet Services and Applications*, vol. 1, no. 1, pp. 7–18, 2010.

[ZHO 09a] ZHOU P., ZHAO B., YANG J. *et al.*, "Energy reduction for STT-RAM using early write termination", *2009 IEEE/ACM International Conference on Computer-Aided Design – Digest of Technical Papers*, pp. 264–268, November 2009.

[ZHO 09b] ZHOU P., ZHAO B., YANG J. *et al.*, "A durable and energy efficient main memory using phase change memory technology", *SIGARCH Computer Architecture News*, vol. 37, no. 3, pp. 14–23, ACM, June 2009.

Index

Printed in the United States
By Bookmasters